Cosmology of Consciousness: Quantum Physics & Neuroscience of Mind

Contents Selected From Volumes 3, 13, and 14
Journal of Cosmology

Editor of Volume 3
Subhash Kak, Ph.D.
Oklahoma State University, Oklahoma

Editor of Volume 13
Rudolf Schild, Ph.D.
Harvard-Smithsonian, Cambridge

Editors of Volume 14
Sir Roger Penrose
University of Oxford, Oxford, United Kingdom
Stuart Hameroff, M.D.
University of Arizona, Arizona,

Cosmology of Consciousness: Quantum Physics & Neuroscience of Mind

Contents Selected From Volumes 3, 13, and 14
Journal of Cosmology

Editor-in-Chief
Rudolf Schild, Ph.D.
Center for Astrophysics,
Harvard-Smithsonian, Cambridge,

Cosmology Science Publishers, Cambridge, 2011
Lana Tao, Managing Editor

Copyright © 2009, 2010, 2011

Published by: Cosmology Science Publishers, Cambridge, MA

All rights reserved. This book is protected by copyright. No part of this book may be reproduced in any form or by any means, including photocopying, or utilized in any information storage and retrieval system without permission of the copyright owner.

The publisher has sought to obtain permission from the copyright owners of all materials reproduced. If any copyright owner has been overlooked please contact: Cosmology Science Publishers at Editor@Cosmology.com, so that permission can be formally obtained.

ISBN: 9780970073358

Key Words: Quantum Physics, Consciousness, Neuroscience, Brain, Mind

Table of Contents

1. How Consciousness Becomes the Physical Universe. Menas Kafatos, Rudolph E. Tanzi, and Deepak Chopra   -5

2. Cosmological Foundations of Consciousness.
Chris King   -17

3. The Origin of the Modern Anthropic Principle.
Helge Kragh   -41

4. Consciousness in the Universe: Neuroscience, Quantum Space-Time Geometry and Orch OR Theory.
Roger Penrose, and Stuart Hameroff   -50

5. What Consciousness Does: A Quantum Cosmology of Mind. Chris J. S. Clarke   -95

6. Quantum Physics & the Multiplicity of Mind:
Split-Brains, Fragmented Minds, Dissociation,
Quantum Consciousness. R. Joseph   -103

7. Logic of Quantum Mechanics and Phenomenon of Consciousness. Michael B. Mensky   -143

8. Evolution of Paleolithic Cosmology and Spiritual Consciousness. R. Joseph   -157

9. Alien Life and Quantum Consciousness, Randy D. Allen   -211

10. Evolution of Consciousness in the Ancient Corners of the Cosmos. R. Joseph   -213

Cosmology of Consciousness

How Consciousness Becomes the Physical Universe

Menas Kafatos, Ph.D.[1], Rudolph E. Tanzi, Ph.D.[2], and Deepak Chopra, M.D.[3]

[1]Fletcher Jones Endowed Professor in Computational Physics, Schmid College of Science, Chapman University, One University Dr. Orange, California, 92866, [2]Joseph P. and Rose F. Kennedy Professor of Neurology, Harvard Medical School Genetics and Aging Research Unit Massachusetts General Hospital/ Harvard Medical School 114 16th Street Charlestown, MA 02129
[3]The Chopra Center for Wellbeing, Carlsbad, CA 92009

Abstract

Issues related to consciousness in general and human mental processes in particular remain the most difficult problem in science. Progress has been made through the development of quantum theory, which, unlike classical physics, assigns a fundamental role to the act of observation. To arrive at the most critical aspects of consciousness, such as its characteristics and whether it plays an active role in the universe requires us to follow hopeful developments in the intersection of quantum theory, biology, neuroscience and the philosophy of mind. Developments in quantum theory aiming to unify all physical processes have opened the door to a profoundly new vision of the cosmos, where observer, observed, and the act of observation are interlocked. This hints at a science of wholeness, going beyond the purely physical emphasis of current science. Studying the universe as a mechanical conglomerate of parts will not solve the problem of consciousness, because in the quantum view, the parts cease to be measureable distinct entities. The interconnectedness of everything is particularly evident in the non-local interactions of the quantum universe. As such, the very large and the very small are also interconnected. Consciousness and matter are not fundamentally distinct but rather are two complementary aspects of one reality, embracing the micro and macro worlds. This approach of starting from wholeness reveals a practical blueprint for addressing consciousness in more scientific terms.

Cosmology of Consciousness

1. Introduction

We realize that the title of our paper is provocative. It is aimed at providing a theory of how the physical universe and conscious observers can be integrated. We will argue that the current state of affairs in addressing the multifaceted issue of consciousness requires such a theory if science is to evolve and encompass the phenomenon of consciousness. Traditionally, the underlying problem of consciousness has been excluded from science, on one of two grounds. Either it is taken as a given that it has no effect on experimental data, or if consciousness must be addressed, it is considered subjective and therefore unreliable as part of the scientific method. Therefore, our challenge is to include consciousness while still remaining within the methods of science.

Our starting point is physics, which recognizes three broad approaches to studying the physical universe: classical, relativistic, and quantum. Classical Newtonian physics is suitable for most everyday applications, yet its epistemology (method of acquiring knowledge) is limited -- it does not apply at the microscopic level and cannot be used for many cosmic processes. Between them, general relativity applies at the large scale of the universe and quantum theory at the microcosmic level. Despite all the attempts to unify general relativity with quantum theory, the goal is still unreached. Of the three broad approaches, quantum theory has clearly opened the door to the issue of consciousness in the measurement process, while relativity admits that observations from different moving frames would yield different values of quantities. Many of the early founders of quantum mechanics held the view that the participatory role of observation is fundamental and the underlying "stuff" of the cosmos is processes rather than the construct of some constant, underlying material substance.

However, quantum theory does not say anything specific about the nature of consciousness -- the whole issue is clouded by basic uncertainty over even how to define consciousness. A firm grasp of human mental processes still remains very elusive. We believe that this indicates a deeper problem which scientists in general are reluctant to address: objective science is based on the dichotomy between subject and object; it rests on the implicit assumption that Nature can be studied ad infinitum as an external objective reality. The role of the observer is, at best, secondary, if not entirely irrelevant.

2. Consciousness and Quantum Theory

In our view, it may well be that the subject-object dichotomy is false to begin with and that consciousness is primary in the cosmos, not just an epiphenomenon of physical processes in a nervous system. Accepting this assumption would turn an exceedingly difficult problem into a very simple one. We will sidestep any precise definition of consciousness, limiting ourselves for now to willful actions on the part of the observer. These actions, of course, are the outcome of specific choices in the mind of the observer. Although some mental actions could be automated, at some point the will of conscious observer(s) sets the whole mechanical aspects of observation in motion.

The issue of observation in QM is central, in the sense that objective reality cannot be disentangled from the act of observation, as the Copenhagen Interpretation (CI) clearly states (cf. Kafatos & Nadeau 2000; Kafatos 2009; Nadeau and Kafatos, 1999; Stapp 1979; Stapp 2004; Stapp 2007). In the words of John A. Wheeler (1981), we live in an observer-participatory universe. The vast majority of today's practicing physicists follow CI's practical prescriptions for quantum phenomena, while still clinging to classical beliefs in observer-independent local, external reality (Kafatos and Nadeau 2000). There is a critical gap between practice and underlying theory. In his Nobel Prize speech of 1932, Werner Heisenberg concluded that the atom "has no immediate and direct physical properties at all." If the universe's basic building block isn't physical, then the same must hold true in some way for the whole. The universe was doing a vanishing act in Heisenberg's day, and it certainly hasn't become more solid since.

This discrepancy between practice and theory must be confronted, because the consequences for the nature of reality are far-reaching (Kafatos and Nadeau, 2000). An impressive body of evidence has been building to suggest that reality is non-local and undivided. Non-locality is already a basic fact of nature, first implied by the Einstein-Podolsky-Rosen thought experiment (EPR, 1935), despite the original intent to refute it, and later explicitly formulated in Bell's Theorem (Bell, 1964) and its relationship to EPR – for further developments, see also experiments which favor QM over local realism, e.g. Aspect, Grangier, and Roger, 1982; Tittel, Brendel, Zbinden & Gisin, 1998. One can also cite the Aharonov-Bohm (1959) effect, and numerous other quantum phenomena.

Moreover, this is a reality where the mindful acts of observation play a crucial role at every level. Heisenberg again: "The atoms or elementary particles themselves . . . form a world of potentialities or possibilities rather than one of things or facts." He was led to a radical conclusion that underlies our own view in this paper: "What we observe is not nature itself, but nature exposed to our method of questioning." Reality, it seems, shifts according to the observer's conscious intent. There is no doubt that the original CI was subjective (Stapp, 2007). However, as Bohr (1934) and Heisenberg (1958) as well as the other developers of CI stated on many occasions, the view that emerged can be summarized as, "the purpose is not to disclose the real essence of phenomena but only to track down... relations between the multifold aspects of our experience" (Bohr, 1934). Stapp (2007) restates this view as "quantum theory is basically about relationships among conscious human experiences" (Stapp 2007). Einstein fought against what he considered the positivistic attitude of CI, which he took as equivalent to Berkeley's dictum to be is to be perceived (Einstein 1951), but he nevertheless admitted that QM is the only successful theory we have that describes our experiences of phenomena in the microcosm.

Quantum theory is not about the nature of reality, even though quantum physicists act as if that is the case. To escape philosophical complications, the original CI was pragmatic: it concerned itself with the epistemology of quantum world (how we experience quantum phenomena), leaving aside ontological questions about the ultimate nature of reality (Kafatos and Nadeau, 2000). The practical bent of CI should be kept in mind, particularly as there is a tendency on the part of many good physicists to slip back into issues that cannot be tested and therefore run counter to the basic tenets of scientific methodology.

To put specifics into the revised or extended CI, Stapp (2007) discusses John von Neumann's different types of processes. The quantum formalism eloquently formalized by von Neumann requires first the acquisition of knowledge about a quantum system (or probing action) as well as a mathematical formalism to describe the evolution of the system to a later time (usually the Schrödinger equation). There are two more processes that Stapp describes: one, according to statistical choices prescribed by QM, yields a specific outcome (or an intervention, a "choice on the part of nature" in Dirac's words); the second, which is primary, preceding even the acquisition of knowledge, involves a "free choice" on the part of the observer. This selection process is not and cannot be described by QM, or for that matter, from any "physically described part of reality" (Stapp, 2007).

These extensions (or clarifications) of the original orthodox CI yield a profoundly different way of looking at the physical universe and our role in it (Kafatos and Nadeau, 2000). Quantum theory today encompasses the interplay of the observer's free choices and nature's "choices" as to what constitute actual outcomes. This dance between the observer and nature gives practical meaning to the concept of the participatory role of the observer. (Henceforth we won't distinguish between the original CI and as it was extended by von Neumann—referring to both as orthodox quantum theory.) As Bohr (1958) emphasized, "freedom of experimentation" opens the floodgates of free will on the part of the observer. Nature responds in the statistical ways described by quantum formalism.

Kafatos and Nadeau (2000) and Nadeau and Kafatos (1999) give extended arguments about these metaphysically-based views of nature. CI points to the limits of physical theories, including itself. If any capriciousness is to be found, it should not be assigned to nature, rather to our mindset about how nature ought to work. As we shall see, there are credible ways to build on quantum formalism and what it suggests about the role of consciousness.

3. Quantum Mechanics and the Brain

It is essential that we avoid the mistake of rooting a physical universe in the physical brain, for both are equally rooted in the non-physical. For practical purposes, this means that the brain must acquire quantum status, just as the atoms that make it up have. The standard assumption in neuroscience is that consciousness is a byproduct of the operation of the human brain. The multitude of processes occurring in the brain covers a vast range of spatio-temporal domains, from the nanoscale to the everyday human scale (e.g. Bernroider and Roy, 2004). Even though they differ on certain issues, a number of scientists accept the applicability of QM at some scales in the brain (cf. Kafatos 2009).

For example, Penrose (1989, 1994) and Hameroff and Penrose (1996) postulate collapses occurring in microtubules induced by quantum gravity. In their view, quantum coherence operates across the entire brain. Stapp (2007) prefers a set of different classical brains that evolve according to the rules of QM, in accordance with the uncertainty principle. He contends that bringing in (the still not developed) quantum gravity needlessly complicates the picture.

In order for an integrative theory to emerge, the next step is to connect the quantum level of activity with higher levels. As a specific example of applying quantum-like processes at mesoscale levels, Roy and Kafatos (1999b) have examined the response and percept domains in the cerebellum. They have built a case that complementarity or quantum-like effects may be operating in brain processes. As is well known, complementarity is a cornerstone of orthodox quantum theory, primarily developed by Niels Bohr. Roy and Kafatos imagine a measurement process with a device that selects only one of the eigenstates of the observable A and rejects all others. This is what is meant by selective measurement in quantum mechanics. It is also called filtration because only one of the eigenstates filters through the process. In attempting to describe both motor function and cognitive activities, Roy and Kafatos (1999a) use statistical distance in setting up a formal Hilbert-space description in the brain, which illustrates our view that quantum formalism may be introduced for brain dynamics.

It is conceivable that the overall biological structures of the brain may require global relationships, which come down processes to global complementarity — every single process is subordinated to the whole. Not just single neurons but massive clusters and networks communicate all but instantaneously. One must also account for the extreme efficiency with which biological organisms operate in a holistic manner, which may only be possible by the use of quantum mechanical formalisms at biological, and neurophysiological relevant scales (cf. Frohlich, 1983; Roy and Kafatos, 2004; Bernroider and Roy, 2005; Davies, 2004, 2005; Stapp, 2004; Hameroff et. al., 2002; Hagan et. al., 2002; Hammeroff and Tuszynski, 2003; Mesquita et. al., 2005; Hunter, 2006; Ceballos et al., 2007).

Stepping into the quantum world doesn't produce easy agreement, naturally. The issue of decoherence (whereby the collapse of the wave function brings a quantum system into relationship with the macro world of large-scale objects and events) is often brought up in arguing against relevant quantum processes in the brain. However, neuronal decoherence processes have only been calculated while assuming that ions, such as K+, are undergoing quantum Brownian motion (e.g. Tegmark, 2000). As such, arguments about decoherence (Tegmark, 2000) assume that the system in question is in thermal equilibrium with its environment, which is not typically the case for bio-molecular dynamics (e.g. Frohlich, 1986; Pokony and Wu, 1998; Mesquita et. al., 2005).

In fact, quantum states can be pumped like a laser, as Frohlich originally proposed for biomolecules (applicable to membrane proteins, and tubulins in microtubules, see also work by Anirban, present volume). Also, experiments and theoretical work indicate that the ions themselves do not move freely within the ion-channel filter, but rather their states are pre-selected, leading to possible protection of quantum coherence within the ion channel for a time scale on the order of 10-3 seconds at 300K, ~ time scale of ion-channel opening and closing (e.g. Bernroider and Roy, 2005). Similar timescales apply to microtubular structures as pointed out by Hameroff and his co-workers. Moreover, progress in the last several years in high-resolution atomic X-ray spectroscopy from MacKinnon's group (Jang et al. 2003) and molecular dynamics simulations (cf. Monroe 2002) have shown that the molecular organization in ion channels allows for "pre-organized" correlations, or ion trappings within the selectivity filter of K+ channels. This occurs with five sets of four carbonyl oxygens acting as filters with the K+ ion, bound by eight oxygens, coordinated electrostatic interactions (Bernroider and Roy 2005). Therefore, quantum entangled states of between two subsystems of the channel filter result.

Beyond the brain, evidence has mounted for quantum coherence in biological systems at high temperatures, whereas in the past coherence was thought to apply to systems near absolute zero. For proteins supporting photosynthesis (Engel, et.al., 2007), solar photons on plant cells are converted to quantum electron states which propagate or travel through the relevant protein by all possible quantum paths, in reaching the part of the cell needed for conversion of energy to chemical energy. As such, new quantum ideas and laboratory evidence applicable to the fields of molecular cell biology and biophysics will have a profound impact in modeling and understanding the process of coherence within neuro-molecular systems.

4. Bridging the Gap: A Consciousness Model

Our purpose here is not to settle these technical issues – or the many others that have arisen as theorists attempt to link quantum processes to the field of biology – but to propose that technical considerations are secondary. What is primary is to have a reliable model against which experiments can offer challenges. Such a model isn't available as long as we fail to account for the disappearance of the material universe implied by quantum theory. This disappearance is real. There is at bottom no strictly mechanistic, physical foundation for the cosmos. The situation is far more radical than most practicing scientists suppose. Whatever is the fundamental source of creation, it itself must be uncreated. Otherwise, there is a hidden creator lying in the background, and then we must ask who or what created that.

What does it mean to be uncreated? The source of reality must be self-sufficient, capable of engendering complex systems on the micro and macro scale, self-regulating, and holistic. Nothing can exist outside its influence. Ultimately, the uncreated source must also turn into the physical universe, not simply oversee it as God or the gods do in conventional religion. We feel that only consciousness fits the bill, for as a prima facie truth, no experience takes place outside consciousness, which means that if there is a reality existing beyond our awareness (counting mathematics and the laws of physics as 1 part of our conscious experience), we will never be able to know it. The fact that consciousness is inseparable from cognition, perception, observation, and measurement is undeniable; therefore, this is the starting point for new insights into the nature of reality.

What is the nature of consciousness in our model? We take it as a field phenomenon, analogous to but preceding the quantum field. This field is characterized by generalized principles already described by quantum physics: complementarity, non-locality, scale-invariance and undivided wholeness. But there is a radical difference between this field and all others: we cannot define it from the outside. To extend Wheeler's reasoning, consciousness includes us human observers. We are part of a feedback loop that links our conscious acts to the conscious response of the field. In keeping with Heisenberg's implication, the universe presents the face that the observer is looking for, and when she looks for a different face, the universe changes its mask.

Cosmology of Consciousness

Consciousness includes human mental processes, but it is not just a human attribute. Existing outside space and time, it was "there" "before" those two words had any meaning. In essence, space and time are conceptual artifacts that sprang from primordial consciousness. The reason that the human mind meshes with nature, mathematics, and the fundamental forces described by physics, is no accident: we mesh because we are a product of the same conceptual expansion by which primordial consciousness turned into the physical world. The difficulty with using basic terms like "concept" and "physical" is that we are accustomed to setting mind apart from matter; therefore, thinking about an atom isn't the same as an atom. Ideas are not substances. But if elementary particles and all matter made of them aren't substances, either, the playing field has been leveled. Quantum theory gives us a model that applies everywhere, not just at the micro level. The real question, then, isn't how to salvage our everyday perception of a solid, tangible world but how to explore the mysterious edge where micro processes are transformed into macro processes, in other words, how Nature gets from microcosm to macrocosm. There, where consciousness acquires the nature of a substance, we must learn how to unify two apparent realities into one. We can begin to tear down walls, integrating objects, events, perceptions, thoughts, and mathematics under the same tent: all can be traced back to the same source.

Physics can serve a pivotal role in transitioning to this new model, because the entire biosphere operates under the same generalized principles we described from the quantum perspective, as does the universe itself. This simple unifying approach must be taken, we realize, as a basic ontological assumption, since it cannot be proven in an objective sense. We cannot extract consciousness from the physical universe, despite the fervent hope of materialists and reductionists. They are forced into a logical paradox, in fact, for either the molecules that make up the brain are inherently conscious (a conclusion to be abhorred in materialism), or a process must be located and described by which those molecules invent consciousness -such a process has not and never will be specified. It amounts to saying that table salt, once it enters the body, finds a way to dissolve in the blood, enter the brain, and in so doing learns to think, feel, and reason.

Our approach, positing consciousness as more fundamental than anything physical, is the most reasonable alternative: Trying to account for mind as arising from physical systems in the end leads (at best) to a claim that mathematics is the underlying "stuff" of the universe (or many universes, if you are of that persuasion). No one from any quarter is proposing a workable material substratum to the universe; therefore, it seems untenable to mount a rearguard defense for materialism itself. As we foresee it, the future development of science will still retain the objectivity of present-day science in a more sophisticated and evolved form. An evolved theory of the role of the observer will be generalized to include physical, biological, and most importantly, awareness aspects of existence. In that sense, we believe the ontology of science will be undivided wholeness at every level. Rather than addressing consciousness from the outside and trying to devise a theory of everything on that basis, a successful Theory Of Everything (TOE) will emerge by taking wholeness as the starting point and fitting the parts into it rather than vice versa. Obviously any TOE must include consciousness as an aspect of "everything," but just as obviously current attempts at a TOE ignore this and have inevitably fallen into ontological traps.

The time has come to escape those traps. An integrated approach will one day prevail. When it does, science will become much stronger and develop to the next levels of understanding Nature, to everyone's lasting benefit.

References

Aharonov, Y., and Bohm, D. (1959) Significance of Electromagnetic Potentials in the Quantum Theory, Phys. Rev., 115, 485-491.

Aspect, A., Grangier, P. and Roger, G. (1982) Experimental Realization of Einstein-Podolsky-Rosen-Bohm Gedankenexperiment: A New Violation of Bell's Inequalities, Phys. Rev. Lett., 49, 91-94.

Bell, J.S. (1964) On the Einstein-Podolsky-Rosen paradox, Physics, 1, 195.

Bernroider, G. and Roy, S. (2004) Quantum-classical correspondence in the brain: Scaling, action distances and predictability behind neural signals. FORMA, 19, 55–68.

Bernroider, G. and Roy, S. (2005) Quantum entanglement of K+ ions, multiple channel states, and the role of noise in the brain. In: Fluctuations and Noise in Biological, Biophysical, and Biomedical Systems III, Stocks, Nigel G.; Abbott, Derek; Morse, Robert P. (Eds.), Proceedings of the SPIE, Volume 5841-29, pp. 205-214.

Bohr, N. (1934) Atomic Theory and the Description of Nature, Cambridge, Cambridge University Press. Bohr, N. (1958) Atomic Physics and Human Knowledge, New York: Wiley.

Ceballos, R., Kafatos, M., Roy, S., and Yang, S., (2007) Quantum mechanical implications for the mind-body issues, Quantum Mind 2007, G. Bemoider (ed)

Univ. Salzburg, July, 2007.

Davies, P. (2004) Does Quantum Mechanics play a non-trivial role in Life? BioSystems, 78, 69–79.

Davies, P. (2005) A Quantum Recipe for Life, Nature, 437, 819.

Einstein, A. (1951) In: Albert Einstein: Philosopher-Physicist, P.A. Schilpp (Ed.) New York, Tudor.

Einstein, A., Podolsky, B., and Rosen, N. (1935) Can Quantum-Mechanical Description of Physical Reality Be Considered Complete?, Phys. Rev., 47, 777-780.

Engel, G.S., Calhoun, T.R., Read, E.L., Ahn, T.K., Mancal, T., Cheng, Y.C., Blankenship, R.E., and Fleming, G.R. (2007) Evidence for wavelike energy transfer through quantum coherence in photosynthetic systems, Nature, 446, 782-786.

Fröhlich, H. (1983) Coherence in Biology, In: Coherent Excitations in Biological Systems, Fröhlich, H. and Kremer, F. (Eds.), Berlin, Springer-Verlag, pp. 1-5.

Fröhlich, H. (1986) In: Modern Bioelectrochemistry, F. Gutman and H. Keyzer (Eds.) Springer-Verlag, New York.

Hagan, S., Hameroff, SR., and Tuszynski, JA. (2002) Quantum computation in brain microtubules: Decoherence and biological feasibility, Physical Review E., 65(6), Art. No. 061901 Part 1 June.

Hameroff, S. and Penrose, R. (1996) Conscious Events as Orchestrated Space-Time Selections, Journal of Consciousness Studies, Vol 3, No. 1, 36-53.

Hameroff, S. and Tuszynski, J. (2003) Search for quantum and classical modes of information processing in microtubules: Implications for the living state, In: Bioenergetic organization in living systems,Eds. Franco Mucumeci, Mae-Wan Ho, World Scientific, Singapore.

Hameroff, S., Nip, A., Porter, M. and Tuszynski, J. (2002) Conduction pathways in microtubules, biological quantum computation, and consciousness, Biosystems, 64(1-3), 149-168.

Heisenberg, W. (1958) Physics and Philosophy, New York, Harper.

Hunter, P. (2006) A quantum leap in biology. One inscrutable field helps another, as quantum physics unravels consciousness, EMBO Rep., October, 7(10), 971–974.

Jiang, Y.A., Lee, A., Chen, J., Ruta, V., Cadene, M., Chait, B.T., and MacKinnon, R. (2003) X-ray structure of a voltage-dependent K+ channel, Nature, 423, 33-41.

Kafatos, M. and Nadeau, R. (1990; 2000). The Conscious Universe: Parts and Wholes in Physical Reality, New York: Springer-Verlag.

Kafatos, M. (2009) Cosmos and Quantum: Frontiers for the Future, Journal of Cosmology, 3, 511-528.

Mesquita, M.V., Vasconcellos, A.R., Luzzi, R., and Mascarenhas, S. (2005) Large-scale Quantum Effects in Biological Systems, Int. Journal of Quantum Chemistry, 102, 1116–1130.

Monroe, C. (2002) Quantum information processing with atoms and photons, Nature, 416, 238-246.

Nadeau, R., and Kafatos, M. (1999) The Non-local Universe: The New Physics and Matters of the Mind, Oxford, Oxford University Press.

Penrose, R. (1989) The Emperor's New Mind, Oxford University Press, Oxford, England. Penrose, R. (1994) Shadows of the Mind, Oxford University Press, Oxford, England.

Pokorny, J., and Wu, T.M. (1998) Biophysical Aspects of Coherence and Biological Order, Springer, New York.

Rosa, L.P., and Faber, J. (2004) Quantum models of the mind: Are they compatible with environment decoherence? Phys. Rev. E, 70, 031902.

Roy S., and Kafatos, M. (1999a) Complemetarity Principle and Cognition Process, Physics Essays, 12, 662-668.

Roy, S., and Kafatos, M., (1999b) Bell-type Correlations and Large Scale Structure of the Universe, In: Instantaneous Action at a Distance in Modern Physics: Pro and Contra, A. E. Chubykalo, V. Pope, & R. Smirnov-Rueda (Eds.), New York: Nova Science Publishers.

Roy, S., and Kafatos, M. (2004) Quantum processes and functional geometry: new perspectives in brain dynamics. FORMA, 19, 69.

Stapp, H.P. (1979) Whiteheadian Approach to Quantum Theory and the Generalized Bell's Theorem, Found. of Physics, 9, 1-25.

Stapp, H. P. (2004) Mind, Matter and Quantum Mechanics (2nd edition), Heidelberg: Springer-Verlag.

Stapp, H.P. (2007) The Mindful Universe: Quantum Mechanics and the Participating Observer, Heidelberg: Springer-Verlag.

Tegmark, M. (2000) Importance of quantum decoherence in brain processes, Phys Rev E Stat Phys Plasmas Fluids Relat Interdiscip Topics, 2000 Apr, 61(4 Pt B), 4194-206.

Tittel, W., Brendel, J., Zbinden, H. and Gisin, N. (1998) Violation of Bell Inequalities by Photons More Than 10km Apart, Phys. Rev. Lett., 81, 3563-3566.

Wheeler, J.A. (1981) Beyond the Black Hole, In: Some Strangeness in the Proportion, H. Woolf (Ed.), Reading, Addison-Wesley Publishing Co.

Cosmological Foundations of Consciousness
Chris King, Ph.D.
Emeritus, Mathematics Department, University of Auckland, New Zealand

Abstract

This chapter explores the cosmological foundations of subjective consciousness in the biological brain, from cosmic-symmetry-breaking, through biogenesis, evolutionary diversification and the emergence of metazoa, to humans, presenting a new evolutionary perspective on the potentialities of quantum interactions in consciousness, and the ultimate relationship of consciousness with cosmology.

1: Introduction: Scope and Design

This overview explores the cosmological foundations of consciousness as evidenced in current research and uses this evidence to present a radical view of what subjective consciousness is, how it evolved, and how it might be supported through quantum processes in the biological brain.

To do full justice to this very broad topic within the confines of the special issue and its planned book edition, I have prepared this paper as a short review article, referring to the full research monograph (King 2011b), as supporting online material, containing all the detailed references, a more complete explanation of the ideas and the ongoing state of the research in the diverse areas covered.

2: Non-linear Quantum and Cosmological Foundations of Biogenesis

While it is well understood that the fundamental forces of nature appear to have differentiated from a super-force in a founding phase of cosmic inflation, the interactive implications of cosmic symmetry-breaking for the chemical basis of life and its evolution into complex sentient organisms are equally as striking, and central to our existence. Cosmic symmetry-breaking and the ensuing preponderance of matter over anti-matter results in the hierarchi cal arrangement of quarks into neutrons and positively charged protons and then the 100 or so stable atomic nuclei, through the interaction of the strong and weak nuclear forces with electromagnetic charge, providing a rich array of stable, electromagnetically polarized, atoms with graduated energetics.

Cosmology of Consciousness

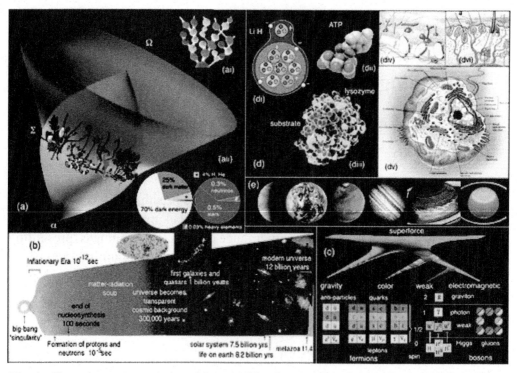

Fig 1: Cosmic symmetry-breaking and its interactive fractal and chaotic effects leading to biogenesis. (a) Life is the consummation of interactive complexity (Σ) resulting from symmetry-breaking of the fundamental force of nature in the big-bang (α), whatever ultimate fate is in store (Ω). Inset (i) fractal inflation model, (ii) the distribution of dark energy and matter and the matter of stars and planets. (b) Logarithmic time scale of cosmological events showing life on earth existing for a third of the universe's lifetime. (c) Symmetry-breaking of the forces of nature results in the color and weak forces, generating 100 atomic nuclei, while gravity and electromagnetism govern long-range structure determining biogenesis, from fractal chemical bonding, to solar systems capable of photosynthetic life in the goldilocks zone of liquid water. (d) Interactive effects of cosmic symmetry-breaking lead to hierarchical interaction of the forces, generating hadrons, atomic nuclei and molecules (i). Non-linear energetics of chemical bonding lead to a cascade of cooperative weak-bonding effects, which generate fractal molecular complexity, from the molecular orbitals of simple molecules (ii), through the 3D structures of complex proteins and nucleic acids (iii) to supra-molecular cell organelles (iv), cells (v), and tissues (vi) and organisms. (e) Chaotic effects of gravity as a non-linear force, results in extreme planetary variation, generating a diversity of potential conditions for biogenesis, similar to dynamic variations surrounding the Mandelbrot set.

The non-linear molecular orbital charge energetics that results in strong covalent and ionic bonds also leads to a cascade of successively weaker bonding effects from H-bonds, to van-der-Waal's interactions, whose globally cooperative nature is responsible for the primary, secondary, and tertiary structures of proteins and nucleic acids, and in a fractal manner to quaternary supra-molecular assemblies, cell organelles, cells, tissues and organisms. Thus, although genetic coding is a necessary condition for the development of cell organelles and organismic tissues, this is possible only because the symmetry-broken laws of nature can give rise to such dynamical structures. In this sense, tissue, culminating in the sentient brain, is the natural interactive full-complexity product of cosmic symmetry-breaking. Despite the periodic quantum properties of the s, p, d and f-orbitals, which form the basis of the table of the elements, successive rows have non-periodic trends because of non-linear charge interactions, which result in a symmetrybreaking determining the bioelements pivotal to biogenesis. Life as we know it is based on the strong covalent bonding of first row elements C, N and O in relation to H, stemming from the optimally strong multiple -CN, -CC-, and >CO bonds, which are cosmically abundant in forming star systems and readily undergo polymerization to heterocyclic molecules, including the nucleic acid bases A, U, G, C and a variety of amino acids, as well as optically active cofactors such as porphyrins.

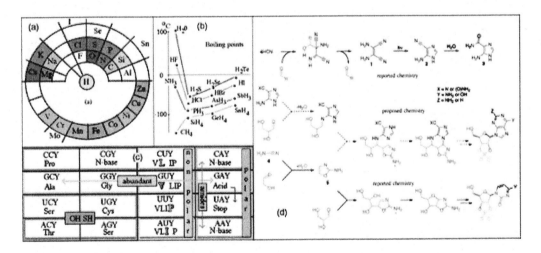

Fig 2: (a) Symmetry-breaking quasi-periodic table of the bioelements displays covalent optimality. (b) Optimality of H20 in terms of internal weak-bonding expressed in its high boiling point. (c) Evidence for a symmetry-breaking origin of the genetic code. (d) Realized and proposed direct synthesis paths from primordial precursors such as HCN to nucleotides (Powner et. al. (2010).

Cosmology of Consciousness

This interactive symmetry-breaking continues in a cascade. As we trend from C > N > O the electronegativity increases from non-polar C-H, to highly electronegative O, resulting in H2O having extreme optimal properties as a polar hydride, bifurcating molecular dynamics into polar and non-polar phases, in addition to pH, and H-bonding effects, which define the aqueous structures and dynamics of proteins, nucleic acids, lipid membranes, ion and electron transport. Following on are secondary properties of S in lower energy -SH and -SS- bonds and the role of P as oligomeric phosphates in the energetics of biogenesis, cellular metabolism, dehydration polymerization and the nucleic acid backbone. We then have bifurcations of ionic properties K+/Na+ and Ca++/Mg++ and finally the catalytic roles of transition elements as trace ingredients.

This does not imply that this is the only elemental arrangement possible for life, as organisms claimed to be adapted to use arsenic in the place of phosphorus (Wolfe-Simon et. al. 2010) suggest, but it does confirm that life as we know it has optimal symmetry-breaking properties cosmologically. Many of the fundamental molecules associ ated with membrane excitation, including lipids such as phosphatidyl choline and amine-based neurotransmitters, also have potentially primordial status (King 1996). Effects of symmetry-breaking may also extend to the genetic code (King 1982). Recent research has begun to elucidate a plausible 'one-pot' rout e (Powner et. al. 2010) from simple cosmically abundant molecules such as HCN and HCHO to the nucleotide units making up RNA, giving our genetic origin a potentially cosmological status. There have also been advances with inducing selected RNAs to self-assemble from precursors and assume catalytic functions (see King 2011b).

3: Emergence of the Excitable Cell: From Universal Common Ancestor to Eucaryotes

Looking back at the universal common ancestor of life, likewise indicates a transition through an era in which RNA functioned as both catalyst and replicator, through the establishment of the genetic code, whose ribosomal protein translation units are still RNA-based, to the eventual emergence of DNA-based life, probably through viral genes (King 2011a). However the genetic picture of cell wall proteins is consistent with independent cellular origins of bacteria and archaea, implying more than one evolution of cellular life from a protected environment conducive to naked nucleotide replication (Russell 2011).

Nevertheless, once the branches of cellular life evolved, excitability based on ion channels and pumps rapidly became universal. It has been reported that as early as 3.3 billion years ago there was a massive genetic expansion, which may have contributed to the genes common to all forms of life facilitated by high

20

levels of horizontal gene transfer, promoted by viruses (Joseph 2011).

Estimates of the adaptive computational power of the collective bacterial and archaean genome (King 2011a) give a presentation rate of new combinations of up to 1030 bits per second, compared with the current fastest computer at about 1017 bit ops per second. Corresponding rates for complex life forms are much lower, around 1017 per second, because they are fewer in total number and have lower reproduction rates and longer generation times. This picture of bit rates coincides closely with the Archaean expansion scenario and suggests that evolution has been a two-phase process of genetic algorithm super-computation, which arrived at a global solution to the notoriously intractable protein-folding problems of the central metabolic and electro-chemical pathways, which are later capitalized on by eukaryotes and metazoa.

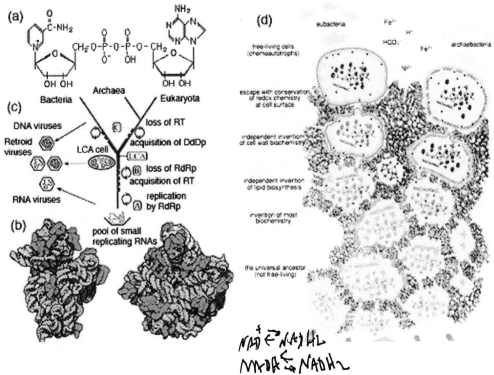

Fig 3: (a) Catalytic nicotine-adenine dinucleotide is essential in respiration. (b) Large and small subunits of the ribosome are centrally and functionally RNA [pink] (c) Molecular fossil evidence for a viral-based cellular transition from the RNA world to DNA based chromosomes, through cellular cooption of viral RNA-directed RNA-polymerase, followed by reverse transcriptase and finally DNA-dependent DNA-polymerase. (d) Independent evolution of archaean and bacterial cellular life from a non-cellular form of life at the interface of olivine and acid, iron-rich sea water forming 'lost city' undersea vents able to solve the concentration and encapsulation problems (Martin and Russell 2003).

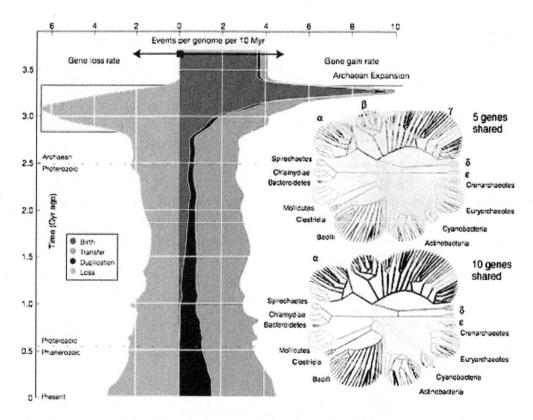

Fig 4: (Left) Archaean genetic expansion around 3.3 billion years ago generated most critical genes common to life (David and Alm 2010) (Right) Evidence of ubiquitous horizontal transfer of genes between bacterial species at different trigger levels (Dagan et. al. 2006).

Horizontal gene transfer, endosymbiosis and gene fusion may have led to a situation where sexuality and excitability, along with all the critical components for neural dynamics including ion-channels specific for $Ca++$, $K+$ and $Na+$, G-protein linked receptors, microtubules, and fast action potential became common among eukaryotes (Wickramasinghe 2011). Ion channel structure appears to have been established during the soup of lateral gene transfers that drove the evolution of eukaryotes. This means we should find neurotransmitter receptors from GABA a, b, and glutamate, through opioid, to dopamine, epinephrine, serotonin and melatonin in all multi-cellular eukaryotes. This universality would have continued up the evolutionary tree, implying that the very different nervous system designs of arthropods and vertebrates mask a deeper common neurodynamic and genetic basis.

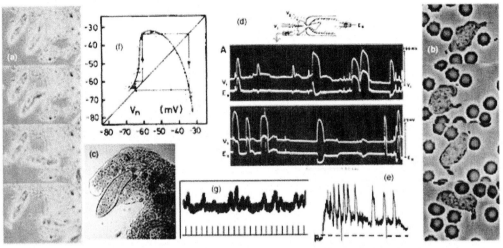

Fig 5: Real-time purposive behavior in single cells (a) Paramecium reverses, turns right and explores a cul-de-sac. (b) Human neutrophil chases an escaping bacterium (black), before engulfing it. (c) Chaos chaos engulfs a paramecium. Action potentials in Chaos chaos (d) and paramecium (e). Period 3 perturbed excitations in alga Nitella indicate chaos. (g) Frog retinal rod cells are sensitive to single quanta in an ultra-low intensity beam.

The evolutionary key to sentient consciousness may lie in the survival advantage it could provide in anticipating threats and strategic opportunities. Since key genes for the brain evolved even before the Cambrian radiation (Wickramasinghe 2011), the key to the emergence of conscious sentience may be sourced in the evolution of excitable single cells. Chaotic excitation provides a eukaryote cell with a generalized quantum sense organ. Sensitive dependence would give a cell feedback about its external environment, perturbed by a variety of quantum modes - chemically through molecular orbital interaction, electromagnetically through photon emission and absorption, electrochemically through the perturbations of the fluctuating fields generated by the excitations, and through acoustic, mechanical and osmotic interaction.

As we move to founding metazoa, we find Hydra, which supports only a primitive diffuse neural net, in continuous transformation and reconstruction, has a rich repertoire of up to 12 forms of 'intuitive' locomotion from snail-like sliding to somersaulting (King 2008), as well as coordinated tentacle movements. This is consistent with much of the adaptive capacity of nervous systems arising from cellular complexity, rather than neural net design alone. Pyramidal neurons for example engage up to 104 synaptic junctions, having a diversity of excitatory and inhibitory synaptic inputs involving up to five types of neurotransmitter, with differing effects depending on receptor types, and their location on dendrites, cell body, or axons.

In the complex central nervous systems of vertebrates, we see the same dynamical features, now expressed in whole system excitations, such as the EEG, in which interacting excitatory and inhibitory neurons provide a basis for broadspectrum oscillation, phase coherence and chaos in the global dynamics, with the synaptic organization enabling the dynamics to resolve complex context-sensitive decision-making problems. Nevertheless the immediate decisionmaking situations around which life or death results, in the theatre of conscious attention are similar to those made by single celled organisms, based strongly on sensory input, and short term anticipation of immediate existential threats and opportunities, in a context of remembered situations that bear upon the current experience.

4: A Dynamical View of the Conscious Brain

Although long distance axons involve pulse coded action potentials, the brain appears to utilize dynamic processing involving broad-spectrum oscillations, rather than discrete signals. Unlike the digital computer, the human brain is a massively parallel organ with perhaps the order of 10 synapses between input and output, despite having an estimated 1010 neurons and 1014 synapses. Such design is essential to enable quick reactions to complex stimuli in real time and avoid the intractability problem of serial computers, which neural nets and genetic algorithms do solve effectively.

The cerebral cortex consists of six layers of cells organized in a sheet of functional columns about 1mm square. These have a fractal modular architecture, with each column representing one aspect of experi ence, from primary processing of lines at given angles, color, motion and auditory tones, through to cells recognizing individual faces. Major areas of the cortex also follow a modular pattern centered on the primary senses and our coordinated motor responses to our ongoing situation. Frontal areas are involved in abstractions of motor events, strategic planning and execution, parietal areas between touch and visual cortices are involved in spatial abstractions, with the temporal lobes extending laterally beyond visual and auditory areas representing attributes with specific meaning, such as specific faces and complex melodies, semantic and symbolic process, such as language, and the temporal relationships between experi ences. This is consistent with a 'holographic' model each experience being represented collectively, like a Fourier transform, in terms of its attributes consistently with the many-to-many connections neurons provide.

No single cortical area has been identified as the seat of consciousness (Joseph

2009). One proposal (Ananthaswamy 2009, 2010) is that conscious processes correspond to the coordinated activity of the whole brain engaging active communication in 'working memory' between the frontal cortex and major sensory and association areas, while activity confined to regional processing is subconscious. This tallies with Bernard Baars' (1997) model of the Cartesian theatre of consciousness as 'global workspace'.

While major input and output pathways pass through thalamic nuclei underlying the cortex, two other systems modulate the dynamics of brain activity. The cortex is energized by ascending pathways from the brain stem, involving the reticular activating system, and dopamine, nor-adrenalin and serotonin pathways, fanning out across wide areas of the cortex, modulating active wakefulness, dreaming and sleep. Our emotional experiences are modulated through the limbic system, a lateral circuit, passing through the hypothalamus regulating internal and hormonal processes, the cingulate cortex dealing with emotional representations, and the hippocampus and amygdala, setting down sequential memories and dealing with flight and fight survival.

There is also evidence active conscious processing corresponds to (30-80 Hz) EEG oscillations in the gamma band, driven by mutual feedback between excitatory and inhibitory neurons in the cortex, and that phase coherence distinguishes 'in-synch' neuronal assemblies forming conscious thought process from peripheral pre-processing (Basar et. al. 1989, Crick & Koch 1992).

While the brain may be 'holographic' spatially, it appears to use phases of dynamical chaos in the time domain. Modulated transitions at the edge of chaos can explain phenomena from perception to insight in a 'eureka' brain wave. In olfactory perception, the brain appears to enter high energy chaos, which frees the dynamic from getting inappropriately locked-in, as annealing does in formal networks, fully-exploring dynamical space, followed by a reduction of energy, causing the dynamic to fall, either into a recognized state, represented by a strange attractor, or to form a new attractor through an adaptive change in the potential energy landscape, through learning (Skarda & Freeman 1987). The same idea fits with the 'eureka' of insight, where an unstable dynamic generated by the problem is resolved in a single bifurcation from chaotic instability into lucidity.

Non-linear mode-locking, common to oscillating chaotic systems, has the potential to facilitate the coherent excitations that characterize coupled neurosystems, going a good way towards resolving the 'binding' problem how the brain 'brings it all back together'. By modulating the coupling between oscillating neurosystems, mode-locking could selectively bring related systems into phase coherence, just as the heartbeat is mode-locked to its local and brain pacemakers.

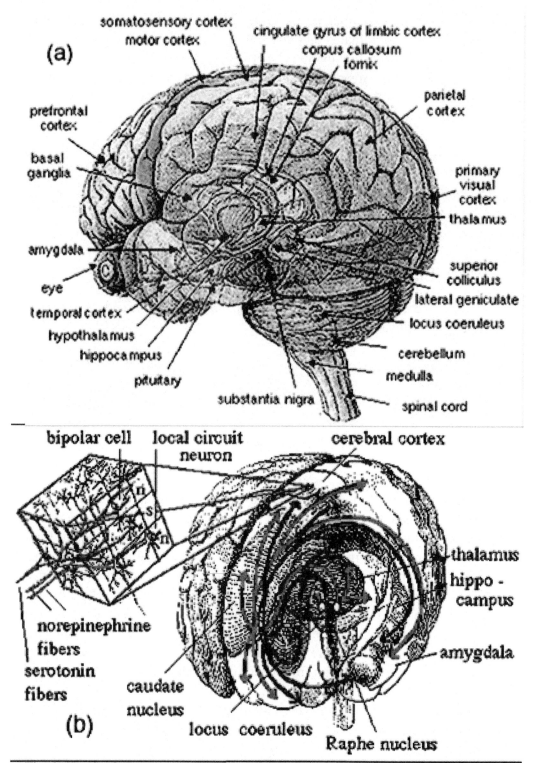

Fig 6a,b,c,d,e,f: Structural overview of the brain as a dynamical organ. (a) Major anatomical features including the cerebral cortex, its underlying driving centres in the thalamus, and surrounding limbic regions involving emotion and memory,

Quantum Physics, Neuroscience of Mind

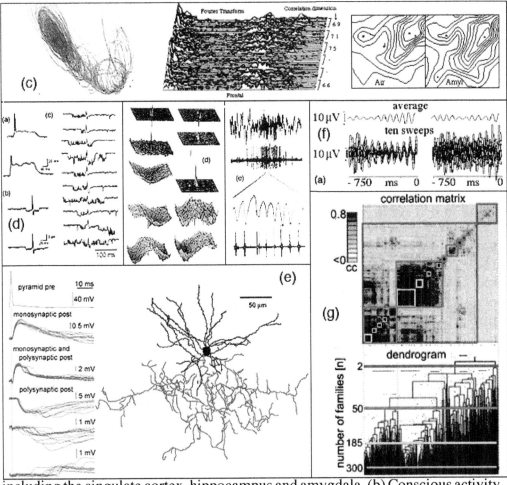

including the cingulate cortex, hippocampus and amygdala. (b) Conscious activity of the cortex is maintained through the activity of ascending pathways from the thalamus and brain stem, including the reticular activating system and serotonin and nor-adrenaline pathways involved in light and dreaming sleep. Processes which enable global dynamics to be affected by small perturbations. (c) Evidence for dynamical chaos includes modulated strange attractors (Freeman 1991), and broad spectrum excitations with moderate fractal (correlation) dimensions (Basar et. al. 1989). These dynamics are complemented by holographic processing across the cortex illustrated in an experimental representation of olfactory excitations corresponding to recognized odors (Skarda and Freeman 1987). (d) Stochastic resonance enables fractal instabilities to grow from ion channel to neuron to hippocampal excitation (Liljenstrm and Uno 2005). (e) Chandelier cells can facilitate an spreading of excitation to many pyramidal cells (Molnar et. al. 2008, Woodruff and Yuste 2008). (f) Wave front coherence in processing becomes manifest when a cue is recognized by the subject (left) (g) Correlation matrix and dendrogram of cortical slice is consistent with fractal self-organized criticality (Beggs and Plenz 2004).

Cosmology of Consciousness

Chaos also makes the brain state arbitrarily sensitive to small perturbations, which is essential for a dynamical brain to be sensitive to small changes in its environment, and to its local instabilities. If the global state is critically poised at a tipping point, an unstable chaotic dynamic could become sensitive to perturbations at the level of the cell, synapse, or ion channel. There are several additional ways in which such sensitivity could come about. Stochastic resonance has been demonstrated to facilitate sensitivity, from ion channel, to cell, to global dynamic (Liljenstrm and Uno 2005). Fractal self-organized criticality has been found in cortical slices (Beggs and Plenz 2004). Chandelier cells have been shown to facilitate lateral spreading of local excitations to multiple pyramidal cells (Molnar et. al. 2008, Woodruff and Yuste 2008).

5: Quantum Dynamics and Conscious Anticipation

The two key questions confounding science about the brain are (1) how and why brain function generates subjective experience, and (2) whether there is any basis for our subjective conscious intentions having physical consequences in 'free-will' (Joseph 2009, 2011).

We thus explore how central to neurodynamic processes might exploit quantum effects to enhance survival prospects of the organism. To develop a realistic quantum theory of consciousness, we have to consider how whole brain states might become capable of quantum interaction (Joseph 2009) and how this could arise from neurophysiological processes common to excitable cells.

We have seen that various forms of global instability, from chaos, through tipping points to self-organized criticality could make the global brain state ultimately sensitive to change at the cellular, molecular or quantum level. Ion channels, such as for acetyl-choline display non-linear (quadratic) concentration dynamics, being excited by two molecules. Many aspects of synaptic release are also highly non-linear, due to biochemical feedback loops. A single vesicle excites up to 2000 ion channels, providing extreme amplification of a potentially quantum event. In addition to being candidates for quantum coherence, voltage gated ion channels display fract al kinetics (Liebovitch 1987).

How interacting systems respond to the quantum suppression of chaos, in processes such as scarring of the wave function (Gutzwiller 1992), received clarification (Chaudhury et. al. 2009, Steck 2009), when it was discovered that an electron in an orbit around a Cs atom in a classically chaotic regime enters into entanglement with nuclear spin. This illustrates how the chaotic 'billiards' of molecular kinetics, and chaotic membrane excitation, might become entangled with other states at the quantum level. One characteristic of time-dependent

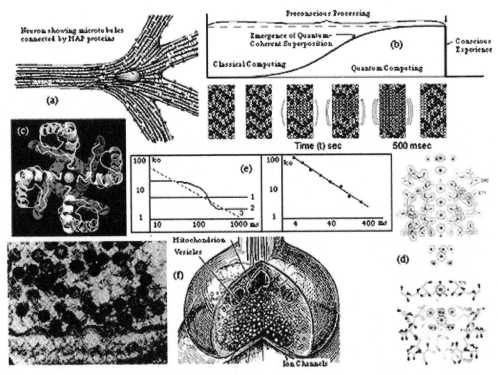

Fig 7: Features of quantum processing in proposed models. (a) Microtubule MAP proteins as envisaged in the OOR model (Hameroff and Penrose 2003). (b) The ensuing relationship between classical and quantum computing and consciousness. (c, d) gated K+ ion channels from MacKinnon's group (Zhou et. al. 2001). (e) Fractal kinetics in the channels (Liebovitch et. al.) (f) Synaptic junction may invoke uncertainty of position of the vesicle.

quantum 'chaos' is transient chaotic behavior ending up in a periodic orbital scar as wave spreading occurs. This would suggest that chaotic sensitivity, with an increasing dominance by quantum uncertainty over time, would contribute to which entanglements ultimately occur in a given kinetic encounter.

The evolutionary argument is a potent discriminator of models of consciousness. Quantum attributes making subjective consciousness possible need to evolve in confluence with essential physiological processes, thus potentially dating back to the epoch when the central components of modulated excitability evolved. Many theories of consciousness have been devised invoking quantum processes which emphasize unusual interpretations of physics, esoteric forms of quantum computation invoking properties extraneous to the known physiological functions of biological organelles, or hypothetical fields in addition to known physiology, raising questions as to whether they pass the evolutionary test. One of the most famous is Hameroff and Penrose's (2003) OOR theory combining objective reduction of the wave function with hypothetical forms of quantum computing on microtubules, which might be extended between cells through gap junctions. These are extensively discussed in the supporting online material, (King 2011b).

One idea fitting closely with neurophysiology is Bernroider's (2003, 2005) proposal that quantum coherence may be sustained in ion channels long enough to be relevant for neural processes and that the channels could be entangl ed with surrounding lipids and proteins and with other channels in the same membrane. He suggests that the ion channel functions through quantum coherence. MacKinnon's group (Zhou et. al. 2001) have shown that the K+-specific ion channel filter works by holding two K+ ions bound to water structures induced by protein side chains. These have similarities to models of quantum computing using ion traps. The solitonic nature of action potentials could provide such entangled connectivity between channels.

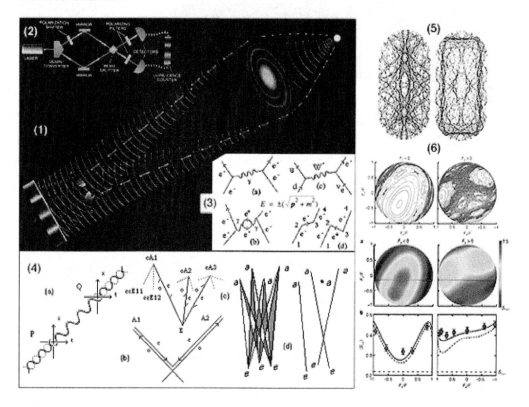

Fig 8: Wheeler delayed choice experiment (1) shows that a decision can be made after a photon from a distant quasar has traversed a gravitationally lensing galaxy by deciding whether to detect which way the photon traveled or to demonstrate it went both ways by sampling interference. The final state at the absorber thus appears to be able to determine past history of the photon. Quantum erasure (2) likewise enables a distinction already made, which would prevent interference,

to be undone after the photon is released. Feynman diagrams (3) show similar time-reversible behavior. In particular time reversed electron scattering (d) is identical to positron creation-annihilation. (4a) In the transactional interpretation (Cramer 1983), a single photon exchanged between emitter and absorber is formed by constructive interference between a retarded offer wave (solid) and an advanced confirmation wave (dotted). (b) Experiments of quantum entanglement involving pair-splitting are resolved by combined offer and confirmation waves, because confirmation waves intersect at the emission point. Contingent absorbers of an emitter in a single passage of a photon (c). Collapse of contingent emitters and absorbers in a transactional match-making (d). (5) Scarring of the wave function of the quantum stadium along repelling orbits (Gutzwiller 1992). (6) Generation of quantum entanglement by quantum chaos in the quantum kicked top (Chaudhury et. al. 2009, Steck 2009).

While decoherence theories and objective reduction do not provide an active role for will, several physicists have suggested consciousness could play a part in the way the wave function representing a superposition of states, collapses to one real instance of the particle. Quantum theory predicts Schrodinger's cat subjected to cyanide if a radioactive scintillation occurs, is in a shadowy superposition - both alive and dead. In our conscious experience of the real world, we find the cat is either alive or dead. This suggests subjective consciousness could play an intervening role within quantum reality, reducing the superabundance of quantum probability multiverses to the historical process we experience. If so, consciousness may have a direct window on the entangled sub-quantum realm (Joseph 2009). We thus explore a model of quantum anticipation, which could extend back to single celled evolution.

Feynman diagrams of quantum interactions show that the quantum interaction is time-reversible. The diagram for electron scattering, when the scattered electron path is time-reversed, becomes positron creation and annihilation. Moreover in real quantum experiments, such as quantum erasure and the Wheeler delayed-choice experiment, it is possible to change how an intervening wave-particle behaves by making different measurements after the wave-particle has passed through the 'apparatus'. All forms of quantum entanglement possess this time-symmetric property. John Cramer (1983) incorporat ed time-symmetry into the 'transactional interpret ation' of quantum mechanics, in which space-time handshaking between the future and past becomes the basis of each real quantum interaction. The emitter of a particle sends out an offer wave forwards and backwards in time, whose energies cancel. The prospective absorbers respond with confirmation waves, and the real quantum exchange arises from constructive interference between the retarded component of the chosen emitter's offer wave and the advanced, time-reversing component of the chosen absorber's confirmation wave. The boundary conditions defining the exchange thus involve both past and future states of the universe. Upon wave function collapse, the exchanged real particle traveling from the emitter to the absorber is identical with its negative energy anti-particle traveling backwards in time.

The transactional interpretation is a heuristic device, which is not essential to the argument, since its predictions coincide, largely, or exclusively with conventional quantum mechanics, but it does highlight future boundary conditions, which could play a part in conscious anticipation. Regardless of the interpretation of quantum mechanics we use, an exchanged particl e has a wave function existing throughout the space-time interval in which it exists, so any process involving collapse of a wave function has boundary conditions extending in principle throughout space-time, involving future prospective absorbers. Advanced entanglement becomes clear in experiments creating two entangled particles (Aspect 1981), where subsequent measurement of the polarization of one photon immediately results in the other having complementary polarization, although neither had a defined polarization beforehand. The only way this correlation can be maintained within quantum reality is through a wave function extending back to the creation event of the pair and forward again in time to the other particle.

If subjective consciousness has a complementary role to brain function, correlated with entangled, quanta emitted and absorbed by the biological brain, it is then correlated with a superposition of possible states in the brain's future, as well as having access to memories of the past. In pair-splitting experiments, the boundary conditions do not permit a classically-causal exploitation. This does not result in a contradiction here, because the brain state is quantum indeterminate and the conscious experience corresponding to the entangled collapse provides an intuitive 'hunch', not a causal deduction.

A possible basis for the emergence of subjective consciousness, which could also be pivotal in explaining the source of free-will, is thus that the excitable cell gained a fundamental form of anticipation of threats to survival. These cells also evolved the ability to perceive strategic opportunities, through anticipatory quantum non-locality induced by chaotic excitation of the cell membrane, in which the cell becomes both an emitter and absorber of its own excitations. Non-locality in space-time is a fundamental quantum property shared by all physical systems, including macroscopic systems with coherent resonance.

The coherent global excitations in the gamma range associated with conscious states, could thus be the 'excitons' in such a quantum model. Unlike quantum computing, which depends on not being disturbed by decoherence caused by interaction with other quanta. Stringent requirements, avoiding decoherence, may not apply to transactions, where real particle exchange occurs even under scattering.

Quantum phenomena abound in biological tissues. Entanglement has been observed in healthy tissues in quantum coherence MRI imaging and bird navigation has been suggested to use entangled electrons. Excitations in photosynthetic antennae have also been shown to perform spatial quantum computing. Enzyme activation energy transition states and synaptic transmission also use quantum tunneling.

By making the organism sensitive to a short envelope of time, extending into the immediate future, as well as the past, subjective consciousness could thus gain an evolutionary advantage, making the organism sensitive to anticipated threats to survival as well as hunting and foraging opportunities. It is these primary needs, guided by the nuances of hunch and familiarity, rather than formal calculations, that the central nervous systems of vertebrates have evolved to successfully handle. Such temporal anticipation need not be of causal efficacy but just provide a small statistical advantage, complemented by computational brain processes associated with learning, which edge-of-chaos wave processing is ideally positioned to do.

These objectives are shared in precisely the same way by single-celled organisms and complex nervous systems. Because of the vastly longer evolutionary time since the Archaean expansion, than the Cambrian metazoan radiation, and the fact that all the components of neuronal excitability were already present when the metazoa emerged, quantum anticipation could have become an evolutionary feature of single celled eukaryot es, before metazoa evolved.

6: Quantum Sensitivity, Sensory Transduction and Subjective Experience

One of the mysteries that distinguish the richness of subjective conscious experience from the colorless logic of electrodynamics is that sensory experiences of vision, sound, smell and touch are richly and qualitatively so different that it is difficult to see how mere variations in neuronal firing organization can give rise to such qualitatively different subjective affects. How is it when dreaming, or in a psychedelic reverie, we can experience ornate visions, hear entrancing music, or smell fragrances as rich, real, intense and qualitatively diverse as those of waking life? Since the senses are actually fundamental quantum modes by which biological organisms can interact with the physical world, this raises the question whether subjective sensory experience is in some way related to the quantum modes by which the physical senses communicate with the world (Joseph 2009). Clearly our senses are sensitive to the quantum level.

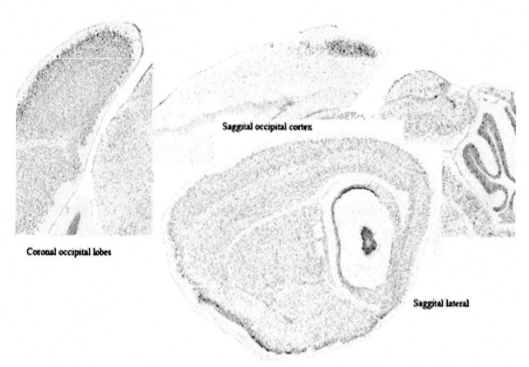

Fig 9: Expression of rhodopsin in the CNS shows both strong selective neuronal activity and a focal expression in the occipital cortex consistent with function in primary visual areas (King 2007).

Individual frog rod cells have been shown to respond to individual photons, the quietest sound involves movements in the inner ear of only the radius of a hydrogen atom and single molecules are sufficient to excite pheromonal receptors. Many genes we associate with peripheral sensory transduction in several senses are also expressed in the mouse brain (King 2007) at least in the form of RNA transcripts, including stomatin-like protein 3 associated with touch, epsin, otocadherin and otoferlin associated with hearing, and several types of opsin, including rhodopsin and encephalopsin. This suggests the brain could harbour an 'internal sensory system' which might play a role in generating the 'internal model of reality', although these ideas are speculative and it is a major challenge to see how such processes could be activated reversibly in the CNS.

Several researchers (Pocket 2000, McFadden 2002) have proposed that neural excitation is associated with electromagnetic fields, which might play a formative role in brain dynamics. Attention has recently been focused on biophotons as a possible basis of processing in the visual cortex based on quantum releases in mitochondrial redox reactions (Rahnama et. al. 2010, Bkkon et. al. 2010). Microtubules have also been implicated (Cifra et. al. 2010).

All excitable cells have ion channels, which undergo conformation changes associated with voltage, and orbital or 'ligand'-binding, both of internal effectors such as G-proteins and externally via neurotransmitters, such as acetylcholine. They also have osmotic and mechano-receptive activation, as in hearing, and in some species can be also activated directly and reversibly by photoreception. Conformation changes of ion channels are capable of exchanging photons, phonons, mechano-osmotic effects and orbital perturbations, representing a form of quantum synesthesia. Since the brain uses up to 40% of our metabolic energy for functions with little or no direct energy output, it is plausible that some of the 'dissipated' energy could be generating novel forms of interaction.

7: Complementarity, Symmetry-breaking, Subjective Consciousness, and Cosmology

This leads to the most perplexing chasm facing the scientific description of reality. What is the existential nature of subjective consciousness, from waking life, through dreaming to psychedelic and mystical experi ence, and does it have cosmological status in relation to the physical universe?

The key entities forming the physical universe manifest as symmetry-broken complementarities. Quanta are waveparticles, with complementary discrete particle and continuous wave aspects. The fundamental forces are symmetry broken in a manner that results in complementary force-radiation bearing bosons and matter forming fermions. In the standard model these have symmetry broken properties, with differing collections of particles. Supersymmetry proposes each boson has a fermion partner to balance their positive and negative energy contributions, but E8's 112 'bosonic' and 128 'fermionic' root vectors, suggest symmetry-breaking could be fundamental (Fielder and King 2010).

Further symmetry-broken complementarities apply to the biological world, where the dyadic sexes of complex organisms and many eukaryot es are both complementary and symmetry broken, with themes of discreteness and continuity even more obviously expressed at the level of sperm and ovum than in our highly symmetry-broken human bodily forms, involving pregnancy, live birth and lactation.

The relationship between subjective consciousness and the physical universe displays a similar complementarity, with profound symmetry breaking. The 'hard problem of consciousness research' (Chalmers 1995) underlines the fundamental di fferences between subjective 'qualia' and the participatory continuity of the Cartesian theatre on the one hand, and the objective, analyzable properties of the physical world around us.

Although we depend on a pragmatic accept ance of the real world, knowing we will pass out if concussed and could die if we cut our veins, from birth to death, the only veridical reality we experience is the envelope of subjective conscious experience. It is only through the consensual regularities of subjective consciousness that we come to know the real world and discover its natural and scientific properties. As pointed out by Indian philosophy, this suggests that mind is more fundamental than matter. The existential status of subjective consciousness thus also has a claim to cosmological status.

A further cosmological interpret ation of consciousness we have noted in association with the cat paradox is that it may function to solve the problem of super-abundance, by reducing probability multiverses to the unique course of history we know and witness. This view of consciousness in shaping the universe is consistent with several of the conclusions of biocentrism (Lanza 2009).

The lessons of quantum and fundamental particle complement arity and symmetry-breaking, sexuality and the Yin-Yang complementarity of the Tao and of Shakti-Shiva in Tantric mind-world cosmologies, lead to a cosmology of consciousness, as symmetry-broken complement to the physical universe.

References

Ananthaswamy, A. (2009) Whole brain is in the grip of consciousness New Scientist 18 March.

Ananthaswamy, A. (2010), Firing on all neurons: Where consciousness comes from, New Scientist, 22 March.

Aspect, A., Grangier P., Roger G. (1981), Phys. Rev. Lett. 47, 460; (1982) Phys. Rev. Lett. 49, 1804; 49, 91.

Baars, B. (1997) In the Theatre of Consciousness: Global Workspace Theory, A Rigorous Scientific Theory of Consciousness. Journal of Consciousness Studies, 4/4 292-309.

Basar E., Basar-Eroglu J., Rschke J., Schtt A., (1989) The EEG is a quasi-deterministic signal anticipating sensory-cognitive tasks, in Basar E., Bullock T.H. eds. Brain Dynamics Springer-Verlag, 43-71.

Beggs J, Plenz D. (2004) Neuronal Avalanches Are Diverse and Precise Activity

Patterns That Are Stable for Many Hours in Cortical Slice Cultures Journal of Neuroscience, 24, 5216-9.

Bernroider, G. (2003) Quantum neurodynamics and the relation to conscious experience Neuroquantology, 2, 1638.

Bernroider, G., Roy, S. (2005) Quantum entanglement of K ions, multiple channel states and the role of noise in the brain SPIE 5841/29 205214.

Bkkon I, Salari V, Tuszynski J, Antal I (2010) Estimation of the number of biophotons involved in the visual perception of a singleobject image: Biophoton intensity can be considerably higher inside cells than outside http://arxiv.org/abs/1012.3371

Chalmers D. (1995) The Puzzle of Conscious Experiencee, Scientific American Dec. 62-69.

Chaudhury S, Smith A, Anderson B, Ghose S, Jessen P (2009) Quantum signatures of chaos in a kicked top, Nature 461 768-771.

Cifra M, Fields J, Farhadi A (2010) Electromagnetic cellular interactions Progress in Biophysics and Molecular Biology doi:10.1016/j.pbiomolbio.2010.07.003

Cramer J.G. (1983) The Transactional Interpretation of Quantum Mechanics, Found. Phys. 13, 887.

Crick F, Koch C. (1992) The Problem of Consciousness, Sci. Am. Sep. 110-117.

Dagan T, Artzy-Randrup Y, Martin W (2006) Modular networks and cumulative impact of lateral transfer in prokaryote genome evolution, PNAS 105/29, 10039-10044.

Darwin C. (1871) The Descent of Man and Selection in Relation to Sex, John Murray, London.

David L, Alm E (2010) Rapid evolutionary innovation during an Archaean genetic expansion Nature doi:10.1038/nature09649
Fielder Christine and King Chris (2004) Sexual Paradox: Complementarity, Reproductive Conflict and Human Emergence Lulu Press.

Gutzwiller, M.C. (1992) Quantum Chaos, Scientific American 266, 78 - 84.

Hameroff, Stuart, Penrose, Roger (2003) Conscious Events as Orchestrated Space-Time Selections, NeuroQuantology; 1, 10-35. Hauser M. (2009) Origin of the Mind, Scientific American, Sept, 44-51.

Joseph, R. (2009). Quantum Physics and the Multiplicity of Mind: Split-Brains, Fragmented Minds, Dissociation, Quantum Consciousness, Journal of Cosmology, 3, 600-640.

Joseph, R. (2011). The neuroanatomy of free will. Loss of will, against the will, "alien hand". Journal of Cosmology, 14, 6000-6045.

King C.C. (1982) A Model for the Development of Genetic Translation, Origins of Life, 12 405-417.

King C.C, (1996) Fractal Neurodynamics and Quantum Chaos : Resolving the Mind-Brain Paradox through Novel Biophysics, in Advances in Consciousness Research, The Secret Symmetry : Fractals of Brain Mind and Consciousness (eds.) E. Mac Cormack and M. Stamenov, John Benjamin.

King C.C. (2007) Sensory Transduction and Subjective Experience Nature Preceedings hdl:10101/npre.2007.1473.1 2009 edition: Activitas Nervosa Superior, 51/1, 45-50. http://www.dhushara.com/lightf/light.htm

King C.C. (2008) The Central Enigma of Consciousness Nature Preceedings hdl:10101/npre.2008.2465.1 2010 edition: http://www.dhushara.com/enigma/enigma.htm

King C. C. (2011a) The Tree of Life: Tangled Roots and Sexy Shoots: Tracing the genetic pathway from the Universal Common Ancestor to Homo sapiens, DNA Decipher J., 1. http://www.dhushara.com/book/unraveltree/unravel.htm

King C. C., (2011b) Cosmological Foundations of Consciousness http://www.dhushara.com/cosfcos/cosfcos2.html

Lanza, Robert and Berman, Bob (2009) Biocentrism: How Life and Consciousness are the Keys to Understanding the True Nature of the Universe, BenBella, ISBN 978-1933771694
Liebovitch L.S., Sullivan J.M., (1987) Fractal analysis of a voltage-dependent potassium channel from cultured mouse hippocampal neurons, Biophys. J., 52, 979-988.

Liljenstrm Hans, Svedin Uno (2005) Micro-Meso-Macro: Addressing Complex

Systems Couplings, Imperial College Press.

Martin, W. and Russell, M. J. (2003) On the origins of cells: a hypothesis for the evolutionary chemoautotrophic transitions from abiotic geochemistry to prokaryotes, and from prokaryotes to nucleated cells, Phil. Trans. R. Soc. Lond. B 358, 59-85.

McFadden J (2002) The Conscious Electromagnetic Information (Cemi) Field Theory: The Hard Problem Made Easy? Journal of Consciousness Studies, 9/8, 45-60. http://www.surrey.ac.uk/qe/pdfs/mcfadden_JCS2002b.pdf

Molnar, G et. al. (2008) Complex Events Initiated by Individual Spikes in the Human Cerebral Cortex, PLOS Biology, 6/9 222. Pockett, Susan (2000) The Nature of Consciousness, ISBN 0595122159.

Powner M., Sutherland J., Szostak J. (2010) Chemoselective Multicomponent One-Pot Assembly of Purine Precursors in Water, J. Am. Chem. Soc., 132, 16677-16688.

Rahnama M, Bkkon I, Tuszynski J, Cifra M, Sardar P, Salari V (2010) Emission of Biophotons and Neural Activity of the Brain, http://arxiv.org/abs/1012.3371

Russell, M. (2011). Origins, abiogenesis, and the search for life. Cosmology Science Publishers, Cambridge.

Skarda C.J., Freeman W.J., (1987) How brains make chaos in order to make sense of the world, Behavioral and Brain Sciences, 10, 161-195.

Steck D (2009) Passage through chaos, Nature, 461, 736-7.

Woodruff, A and Yuste R (2008) Of Mice and Men, and Chandeliers, PLOS Biology, 6/9, 243.

Zhou, Y., Morais-Cabral, A., Kaufman, A. & MacKinnon, R. (2001) Chemistry of ion coordination and hydration revealed in K+ channel-Fab complex at 2.0 A resolution, Nature, 414, 43-48.

Cosmology of Consciousness

The Origin of the Modern Anthropic Principle

Helge Kragh, Ph.D.

Department of Science Studies, Building 1110, Aarhus University,
Aarhus, Denmark.

Abstract

Since its origin in the early 1970s the anthropic principle has exerted a major influence on ideas of theoretical cosmology. Although it is today as controversial as ever, its impact is beyond discussion. This paper examines some of the early formulations of anthropic ideas, including those of Russian cosmologists G. Idlis and A. Zelmanov. These early formulations are at best vague anticipations of the anthropic principle, which as a research tool for cosmological theory was first proposed by B. Carter in 1973. The paper offers an account of how Carter came to the idea of the anthropic principle and how he originally formulated it.

1. Some Pre-Carter Anthropic Ideas

The general meaning of the anthropic principle is that what we observe must be compatible with our existence or, more generally, with the existence of advanced life. Humans can occupy only a universe like ours, and this explains in a sense why the universe is as it is. Today the anthropic principle is often seen as a selection principle operating in the context of the multiverse, a view which goes back to Brandon Carter's original formulation nearly 40 years ago. The principle has a rich prehistory and anticipations of it can, if only with a considerable amount of hindsight, be found as far back in time as in ancient Greece (Ćirković, 2003; Barrow & Tipler, 1986). However, I shall limit my remarks to a few possible and not well known precursors of the twentieth century, including James Jeans and Arthur Eddington from the period between the two world wars.

Speculations about the role of life in the universe were a common theme in Jeans's many popular addresses. In a lecture of 1926 he argued that the liquid state, and hence the existence of ordinary water, required very special conditions. More generally he emphasized that "the physical conditions under which life is possible form only a tiny fraction of the range of physical conditions which prevail in the universe as a whole." Moreover, "In every respect space, time, physical conditions life is limited to an almost inconceivably small corner of the universe" (Jeans, 1926, p. 40). The fine-tuning necessary for life formed an important part of the later anthropic principle, but Jeans did not explicitly associate it with either the constants of nature or cosmic evolution. Indeed, by 1926 the expanding universe was still in the future.

Eddington was convinced that the laws of nature are indirectly imposed by the human mind which largely determines the nature and extent of what we know about the universe. In his method known as selective subjectivism he appealed to selection arguments somewhat similar to those later associated with the anthropic principle. He argued that the number of particles in the universe ($N=10^{79}$) and most other constants were determined by mental and therefore human factors, namely "the influence of the sensory equipment with which we formulate the results of observation of knowledge." This influence, he said, "is so far-reaching that by itself it decides the number of particles into which the matter of the universe appears to be divided" (Eddington 1939, p. 60). Eddington's anthropic-like reasoning related specifically to the human mind, not to advanced life in general.

It is often stated that Fred Hoyle applied anthropic arguments as early as 1953, namely in his famous prediction of a 7.7 MeV resonance level in carbon-12. In 1952 Edwin Salpeter had suggested a "triple alpha" process according to which three alpha particles would fuse into a carbon-12 nucleus under the physical conditions governing the interior of some stars. Hoyle recognized that Salpeter's process would only work if there existed a resonance level of about 7.7 MeV, which was subsequently found experimentally (Salpeter, 1955; Kragh, 2010). Since carbon is a prerequisite for life as we know it, we would not be here had it not been because of this particular energy level. Hoyle's successful prediction has often been quoted as "the only genuine anthropic prediction" and "evidence to support the argument that the Universe has been designed for our benefit tailor-made for man" (Gribbin & Rees, 1989, p. 247). However, Hoyle does not qualify as the originator of the anthropic principle. Although he did indeed predict the resonance level necessary for the production of carbon, and then life, his prediction owed little to anthropic reasoning (for details and references, see Kragh, 2010).

A few years after Hoyle's non-anthropic prediction, in 1958, the Russian astronomer Grigory Moiseevich Idlis published a paper in the Proceedings of the Astrophysical Institute of the Kazach Academy of Sciences in which he vaguely anticipated the anthropic principle. It is however unclear to what extent, since the obscure paper only exists in Russian language. In a later review Idlis (2001) claimed to be the true discoverer of the anthropic principle, a claim which was supported by his famous compatriot, the pioneer cosmologist Yakov Zel'dovich

(1981). According to Idlis (1982, p. 357), in 1958 he argued that "properties of the Metagalaxy..." universe " are, generally speaking, just the necessary and sufficient conditions for the natural origination and evolution of life to higher intelligence forms of matter, similar to man, finally aware of itself."

If Idlis's paternity to the anthropic principle is questionable, so is the claim that a version of the principle was formulated even earlier by Abraham Leonidovich Zelmanov, a Russian physicist and cosmologist who in the 1940s did important work on inhomogeneous cosmological models (Zelmanov, 2006). Zelmanov argued for an intimate interdependence between observers and the observed universe, including the idealist claim that "If no observers exist then the observable universe as well does not exist." He also stressed that "humanity can exist only with the specific scale of the numerical values of the cosmological constants [and] is only an episode in the life of the universe" (Rabounski, 2006, p. 35). Zelmanov seems never to have published his anthropic considerations. Moreover, what we know of them lacks proper documentation. I believe the priority claims of Idlis and Zelmanov are unconvincing.

2. Large Cosmic Numbers: Dirac and Dicke

While Carter was unaware of Idlis and Zelmanov, he was to some extent influenced by the views of Paul Dirac, Robert Dicke and John Wheeler. The cosmological theory that Dirac proposed in 1937 was based on the relationship between two large dimensionless numbers:

$$\frac{e^2}{GmM} \cong \frac{T_0}{\Delta t} \cong 10^{39}$$

T_0 denotes the Hubble time, M the mass of the proton, and $\Delta t = e^2/mc^3 = 10^{-24}$ s is an atomic time unit; the other symbols have their usual meanings. According to Dirac (1937), the numerical relationship was a permanent feature, valid for any value of the Hubble time and not only for the present one. From this he inferred that the gravitational constant slowly decreased in time according to $G \sim 1/t$. He did not refer to life in his paper of 1937. More than twenty years later Dicke reconsidered the significance of Dirac's large numbers which he expressed in a somewhat different way:

$$N_1 = \frac{\hbar c}{GM^2} \cong 2 \times 10^{38} \quad \text{and} \quad N_2 = \frac{T}{\hbar / Mc^2} \cong 10^{42}$$

where T is the age of the universe, $T \approx T_0$. In terms of the Planck mass the first number, which is equal to the inverse of the gravitational coupling constant, can be written as

$$N_1 = \left(\frac{m_{Pl}}{M} \right)^2 \quad \text{with} \quad m_{Pl} = \sqrt{\frac{ch}{G}}$$

Dicke pointed out that the present age is not random, as assumed by Dirac, but characterized by the existence of carbon and other heavy elements. He argued that "T is not permitted to take one of an enormous range of values, but is somehow limited by the biological requirements to be met during the epoch of man" (Dicke, 1961, p. 441). Dicke's critical analysis of the large numbers found in nature was followed by a reply from Dirac (1961), who knew of the paper in advance of its publication:

On this [Dicke's], assumption habitable planets could exist only for a limited period of time. With my assumption they could exist indefinitely in the future and life need never end. There is no decisive argument for deciding between these assumptions. I prefer the one that allows the possibility of endless life.

Dirac had for a long time been devoted to the doctrine of eternal intelligent life in the universe, a doctrine that would later form the basis of the final anthropic principle (FAP). In private notes of 1933 he stated as his credo that "the human race will continue to live for ever and will develop and progress without limit." What he characterized as an "article of faith" was "an assumption that I must make for my peace of mind. Living is worthwhile only if one can contribute in some small way to this endless chain of progress" (Farmelo, 2009, p. 221).

3. Towards the Anthropic Principle

Australian-born Brandon Carter wanted to understand the role of microphysical parameters in cosmology. In 1967, while a PhD student at Cambridge University, he wrote a manuscript on the subject which circulated among a small number of physicists. Carter's purpose was "to clarify the significance of the famous coincidence between the Hubble age of the universe and a certain combination of microphysical parameters," that is, the same relationship $N_1 = N_2$ which Dicke had considered in 1961. This coincidence, he said, "can be fully explained in principle in terms of conventional physics and cosmology, so that revolutionary departures such as Dirac's hypothesis of varying gravitational constant, or Eddington's Fundamental Theory are not justified" (Carter, 2007, p. 1). Surprisingly, he was at that time unacquainted with the works of Dicke, which he only came to know about the following year (E-mail from Carter to the author, 18 February

2010). Although he was aware of Dirac's Large Number Hypothesis, which he knew from Hermann Bondi's textbook Cosmology, he did not mention it in his manuscript of 1967. It is also worth noticing that he was not yet acquainted with the works of Salpeter and Hoyle on the triple-alpha process generating carbon-12 in the stars.

The first time Carter had an opportunity to present his ideas of what would become the anthropic principle was at a meeting in February 1970 in Princeton organized by Wheeler. At this meeting, a commemoration of the works of the English nineteenth-century mathematician William Kingdon Clifford, he gave a talk on "Large Numbers in Astrophysics and Cosmology" in which he argued that the universe could only be understood if life and observers were taken into account. He did not, at that time, speak of an anthropic principle. Carter's ideas were known to a small group of physicists and cosmologists before they finally appeared in print in 1974. Wheeler, Stephen Hawking and Freeman Dyson had participated in the Clifford memorial meeting in Princeton and discussed Carter's presentation. They were receptive to his ideas, which they found to be fascinating, such as shown by their published responses (Rees et al. 1974, p. 307; Hawking, 1974; Dyson, 1972). As early as 1972, in a review of the possible time variation of the constants of nature, Dyson referred to Carter's "principle of cognizability." Dyson considered Dicke's numbers N_1 and N_2, which he rewrote as

$$ \gamma = \frac{Gm^2}{\hbar c} \quad \text{and} \quad \delta = \frac{\hbar H}{mc^2} $$

where H is the Hubble parameter, $H = 1/T$. He took the principle of cognizability to imply that "the presence in the universe of conscious observers places limits on the absolute magnitudes of γ and δ and not only on their ratio" (Dyson, 1972, p. 235).

4. Carter's Address of 1973

On 10-12 September 1973 the International Astronomical Union held a meeting in Cracow, Poland, dedicated to the 500th anniversary of the birth of Copernicus. One of the sessions was chaired by Wheeler, on whose suggestion Carter gave a presentation of his anthropic considerations (Longair, 1974, p. 289). An extended version of the lecture appeared the following year in the proceedings of the conference edited by the Cambridge astrophysicist Malcolm Longair. The lecture and the corresponding article entitled "Large Number Coincidences and the Anthropic Principle in Cosmology" marks the true beginning of the anthropic principle. Hawking was also present at the Cracow meeting, where he gave a talk in which he suggested that the isotropy of the universe followed from anthropic arguments. He did not refer to the anthropic principle, but to the "Dicke-Carter idea" (Hawking, 1974).

Appropriately, at a meeting celebrating Copernicus, Carter introduced his topic with a reference to the so-called Copernican principle, the doctrine that we do not occupy a privileged position in the universe. But it was a critical reference: "Unfortunately there has been a strong (not always subconscious) tendency to extend this to a most questionable dogma to the effect that our situation cannot be privileged in any sense" (Carter, 1974, p. 291). Carter objected that there is indeed a sense in which our situation is privileged, namely a temporal sense. We obviously live in the epoch of life, which is far from random.

The large number coincidences considered by Eddington, Dirac and Dicke should be understood in terms of "what may be termed the anthropic principle to the effect that what we can expect to observe must be restricted by the conditions necessary for our presence as observers." It is necessary to take into account our special situation and properties when interpreting observational data, just as it is necessary to take into account the special properties of measuring instruments.

The name that Carter introduced in 1974 has proved highly successful, but it is generally recognized, and admitted by Carter himself, that it invites associations that are both unfortunate and unintended. At a conference in 1989 he proposed as a more informative and less ethnocentric name "observer self selection principle," thereby stressing that what is important to most applications of the anthropic principle is observers and not human beings (Carter, 1993, p. 38). Neither this name nor other alternative names, such as the "biophilic" principle, have caught on.

In addition to this "weak" anthropic principle (WAP), Carter introduced a "strong" version (SAP) according to which "the Universe (and hence the fundamental parameters on which it depends) must be such as to admit the creation of observers within it at some stage." Whether in one form or other he thought that the new antropic principle would only have explanatory power if associated with the idea of a world ensemble, the assumption that there exists numerous other universes. These other universes would be really existing and characterized by all possible combinations of initial conditions and fundamental constants. However, contrary to most later ideas of the multiverse, he only considered as real those universes which can accommodate observers of some kind. Carter felt that his suggestion of an anthropic multiverse (a word not yet coined) was justified by Hugh Everett's many-worlds interpretation of quantum mechanics to which "one is virtually forced by the internal logic of quantum theory" (Carter, 1974, p. 298).

At the time this was a view shared only by a minority of physicists. According to Carter's argument our universe is special by providing conditions for intelligent life. Since no reasons can be given for the non-existence of numerous other universes without observers, he found it natural to assume that they do exist.

Carter was well aware that the new anthropic approach to cosmology was unorthodox and problematic from both a scientific and philosophical perspective. Realizing the speculative and non-testable nature of the strong principle he called it "rather more questionable" than the weak one. Did anthropic explanations, relying as they did on the assumption of other universes, qualify as bona fide explanations? "I would personally be happier," he admitted at the end of his paper, "with explanations of the values of the fundamental coupling constants etc. based on a deeper mathematical structure in which they would no longer be fundamental but would be derived." He reckoned a similar kind of problem in relation to anthropic predictions, which differed fundamentally from those ordinarily used in physics. Predictions based on the anthropic principle, he wrote, "will not be completely satisfying from a physicist's point of view since the possibility will remain of finding a deeper underlying theory explaining the relationships that have been predicted." Indeed, over the next decades questions of this kind would be hotly debated, both by cosmologists and philosophers. Currently they are mostly discussed within the framework of the string-based landscape multiverse (Carr, 2007), but they have their origin in Carter's address of 1973.

REFERENCES

Barrow, J. D., Tipler, F. (1986). The Anthropic Cosmological Principle. Oxford University Press, Oxford.

Carr, B. (Ed.) (2007). Universe or Multiverse? Cambridge University Press,

Carter, B. (1974). Large number coincidences and the anthropic principle in cosmology. In: Longair, M. (Ed.), Confrontation of Cosmological Theories with Observational Data. Reidel, Dordrecht, pp. 291-298.

Carter, B. (1993). The anthropic selection principle and the ultra-Darwinian synthesis. In: Bertola, F., Curi, U. (Eds), The Anthropic Principle. Cambridge University Press, Cambridge.

Carter, B. (2007). The significance of numerical coincidences in nature. ArXiv:0710.3543.

Ćirković, M. (2003). Ancient origins of a modern anthropic cosmological argument. Astronomical and Astrophysical Transactions, 22, 879-886.

Dicke, R. (1961). Dirac's cosmology and Mach's principle. Nature, 192, 440-441.

Dirac, P. (1937). The cosmological constants. Nature, 139, 323.

Dirac, P. (1961). Reply to R. H. Dicke. Nature, 192, 441.

Dyson, F. (1972). The fundamental constants and their time variation. In: Salam, A., Wigner, E. P. (Eds), Aspects of Quantum Theory. Cambridge University Press, Cambridge, pp. 213-236.

Eddington, A. S. (1939). The Philosophy of Physical Science. Cambridge University Press, Cambridge.

Farmelo, G. (2009). The Strangest Man: The Hidden Life of Paul Dirac, Quantum Genius. Faber and Faber, London.

Gribbin, J., Rees, M. (1989). Cosmic Coincidences: Dark Matter, Mankind, and Anthropic Cosmology. Bantam Books, New York.

Hawking, S. (1974). The anisotropy of the universe at large times. In: Longair, M. (Ed.), Confrontation of Cosmological Theories with Observational data. Reidel, Dordrecht, pp. 283-286.

Idlis, G. (1982). Four revolutions in astronomy, cosmology and physics. Acta Historiae Rerum Naturalium Nec Non Technicarum, 18, 343-368.

Idlis, G. (2001). Universality of space civilizations and indispensible universality in cosmology. Astronomical and Astrophysical Transactions, 20, 963-973.

Jeans, J. (1926). Recent developments of cosmical physics. Nature, 118, 29-40.

Kragh, H. (2010). An anthropic myth: Fred Hoyle's carbon-12 resonance level. Archive for History of Exact Sciences, 64, 721-751.

Longair, M. (Ed.). Confrontation of Cosmological Theories with Observational Data. Reidel, Dordrecht.

Rabounski, D. (2006). Zelmanov's anthropic principle and the infinite relativity principle. Progress in Physics, 1, 35-37.

Rees, M., Ruffini, R., Wheeler, J. (1998). Black Holes, Gravitational Waves and Cosmology: An Introduction to Current Research. Gordon and Breach, New York.

Salpeter, E. (1955). The 7.68-MeV level in C-12 and stellar energy production. Physical Review, 98, 1183-1184.

Zel'dovich, Y. (1981). The birth of a closed universe, and the anthropogenic principle. Soviet Astronomy Letters, 7, 322-323.

Zelmanov, A. (2006). Chronometric Invariants: On Deformations and the Curvature of Accompanying Space. American Research Press, Rebohoth, NM.

Consciousness in the Universe: Neuroscience, Quantum Space-Time Geometry and Orch OR Theory

Roger Penrose, PhD, OM, FRS[1], and Stuart Hameroff, MD[2]

[1]Emeritus Rouse Ball Professor, Mathematical Institute, Emeritus Fellow, Wadham College,
University of Oxford, Oxford, UK
[2]Professor, Anesthesiology and Psychology, Director, Center for Consciousness Studies, The University of Arizona, Tucson, Arizona, USA

Abstract

The nature of consciousness, its occurrence in the brain, and its ultimate place in the universe are unknown. We proposed in the mid 1990's that consciousness depends on biologically 'orchestrated' quantum computations in collections of microtubules within brain neurons, that these quantum computations correlate with and regulate neuronal activity, and that the continuous Schrödinger evolution of each quantum computation terminates in accordance with the specific Diósi-Penrose (DP) scheme of 'objective reduction' of the quantum state (OR). This orchestrated **OR** activity (Orch OR) is taken to result in a moment of conscious awareness and/or choice. This particular (DP) form of OR is taken to be a quantum-gravity process related to the fundamentals of spacetime geometry, so Orch OR suggests a connection between brain biomolecular processes and fine-scale structure of the universe. Here we review and update Orch OR in light of criticisms and developments in quantum biology, neuroscience, physics and cosmology. We conclude that consciousness plays an intrinsic role in the universe.

1. Introduction: Consciousness, Brain and Evolution

Consciousness implies awareness: subjective experience of internal and external phenomenal worlds. Consciousness is central also to understanding, meaning and volitional choice with the experience of free will. Our views of reality, of the universe, of ourselves depend on consciousness. Consciousness defines our existence.

Three general possibilities regarding the origin and place of consciousness in the universe have been commonly expressed.

(A) *Consciousness is not an independent quality but arose as a natural evolutionary consequence of the biological adaptation of brains and nervous systems.* The most popular scientific view is that consciousness emerged as a property of complex biological computation during the course of evolution. Opinions vary as to when, where and how consciousness appeared, e.g. only recently in humans, or earlier in lower organisms. Consciousness as evolutionary adaptation is commonly assumed to be epiphenomenal (i.e. a secondary effect without independent influence), though it is frequently argued to confer beneficial advantages to conscious species (Dennett, 1991; 1995; Wegner, 2002).

(B) *Consciousness is a quality that has always been in the universe.* Spiritual and religious approaches assume consciousness has been in the universe all along, e.g. as the 'ground of being', 'creator' or component of an omnipresent 'God'. Panpsychists attribute consciousness to all matter. Idealists contend consciousness is all that exists, the material world an illusion (Kant, 1781).

(C) *Precursors of consciousness have always been in the universe; biology evolved a mechanism to convert conscious precursors to actual consciousness.* This is the view implied by Whitehead (1929; 1933) and taken in the Penrose-Hameroff theory of 'orchestrated objective reduction' ('Orch OR'). Precursors of consciousness, presumably with proto-experiential qualities, are proposed to exist as the potential ingredients of actual consciousness, the physical basis of these proto-conscious elements not necessarily being part of our current theories of the laws of the universe (Penrose and Hameroff, 1995; Hameroff and Penrose, 1996a; 1996b).

2. Ideas for how consciousness arises from brain action

How does the brain produce consciousness? An enormous amount of detailed knowledge about brain function has accrued; however the mechanism by which the brain produces consciousness remains mysterious (Koch, 2004). The prevalent scientific view is that consciousness somehow emerges from complex computation among simple neurons which each receive and integrate synaptic inputs to a threshold for bit-like firing. The brain as a network of 10^{11} 'integrate-and-fire' neurons computing by bit-like firing and variable-strength chemical synapses is the standard model for computer simulations of brain function, e.g. in the field of artificial intelligence ('AI').

The brain-as-computer view can account for non-conscious cognitive functions including much of our mental processing and control of behavior. Such non-conscious cognitive processes are deemed 'zombie modes', 'auto-pilot', or 'easy problems'. The 'hard problem' (Chalmers, 1996) is the question of how cognitive processes are accompanied or driven by phenomenal conscious experience and subjective feelings, referred to by philosophers as 'qualia'. Other issues also suggest the brain-as-computer view may be incomplete, and that other approaches are required. The conventional brain-as-computer view fails to account for:

The 'hard problem' Distinctions between conscious and non-conscious processes are not addressed; consciousness is assumed to emerge at a critical level (neither specified nor testable) of computational complexity mediating otherwise non-conscious processes.

'Non-computable' thought and understanding, e.g. as shown by Gödel's theorem (Penrose, 1989; 1994).

'Binding and synchrony', the problem of how disparate neuronal activities are bound into unified conscious experience, and how neuronal synchrony, e.g. gamma synchrony EEG (30 to 90 Hz), the best measurable correlate of consciousness does not derive from neuronal firings.

Causal efficacy of consciousness and any semblance of free will. Because measurable brain activity corresponding to a stimulus often occurs after we've responded (seemingly consciously) to that stimulus, the brain-as-computer view depicts consciousness as epiphenomenal illusion (Dennett, 1991; 1995; Wegner, 2002).

Cognitive behaviors of single cell organisms. Protozoans like Paramecium can swim, find food and mates, learn, remember and have sex, all without synaptic computation (Sherrington, 1957).

In the 1980s Penrose and Hameroff (separately) began to address these issues, each against the grain of mainstream views.

3. Microtubules as Biomolecular Computers

Hameroff had been intrigued by seemingly intelligent, organized activities inside cells, accomplished by protein polymers called microtubules (Hameroff and Watt, 1982; Hameroff, 1987). Major components of the cell's structural cytoskeleton, microtubules also accounted for precise separation of chromosomes in cell division, complex behavior of *Paramecium*, and regulation of synapses within brain neurons (Figure 1). The intelligent function and periodic lattice structure of microtubules suggested they might function as some type of biomolecular computer.

Microtubules are self-assembling polymers of the peanut-shaped protein dimer tubulin, each tubulin dimer (110,000 atomic mass units) being composed of an alpha and beta monomer (Figure 2). Thirteen linear tubulin chains ('protofilaments') align side-to-side to form hollow microtubule cylinders (25 nanometers diameter) with two types of hexagonal lattices. The A-lattice has multiple winding patterns which intersect on protofilaments at specific intervals matching the Fibonacci series found widely in nature and possessing a helical symmetry (Section 9), suggestively sympathetic to large-scale quantum processes.

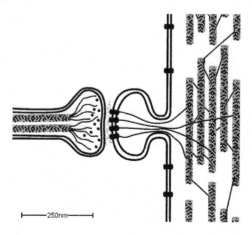

Figure 1. Schematic of portions of two neurons. A terminal axon (left) forms a synapse with a dendritic spine of a second neuron (right). Interiors of both neurons show cytoskeletal structures including microtubules, actin and microtubule-associated proteins (MAPs). Dendritic microtubules are arrayed in mixed polarity local networks, interconnected by MAPs. Synaptic inputs are conveyed to dendritic microtubules by ion flux, actin filaments, second messengers (e.g. CaMKII, see Hameroff et al, 2010) and MAPs.

Along with actin and other cytoskeletal structures, microtubules establish cell shape, direct growth and organize function of cells including brain neurons. Various types of microtubule-associated proteins ('MAPs') bind at specific lattice sites and bridge to other microtubules, defining cell architecture like girders and beams in a building. One such MAP is tau, whose displacement from microtubules results in neurofibrillary tangles and the cognitive dysfunction of Alzheimer's disease (Brunden et al, 2011). Motor proteins (dynein, kinesin) move rapidly along microtubules, transporting cargo molecules to specific locations.

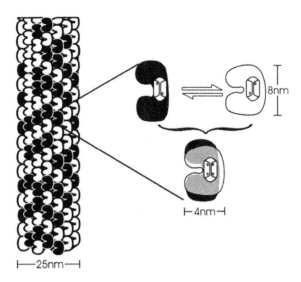

Figure 2. Left: Portion of single microtubule composed of tubulin dimer proteins (black and white) in A-lattice configuration. Right, top: According to pre-Orch OR microtubule automata theory (e.g. Hameroff and Watt, 1982; Rasmussen et al, 1990), each tubulin in a microtubule lattice switches between alternate (black and white) 'bit' states, coupled to electron cloud dipole London forces in internal hydrophobic pocket. Right, bottom: According to Orch OR, each tubulin can also exist as quantum superposition (quantum bit, or 'qubit') of both states, coupled to superposition of London force dipoles in hydrophobic pocket.

Microtubules also fuse side-by-side in doublets or triplets. Nine such doublets or triplets then align to form barrel-shaped mega-cylinders called cilia, flagella and centrioles, organelles responsible for locomotion, sensation and cell division. Either individually or in these larger arrays, microtubules are responsible for cellular and intra-cellular movements requiring intelligent spatiotemporal organization. Microtubules have a lattice structure comparable to computational systems. Could microtubules process information?

The notion that microtubules process information was suggested in general terms by Sherrington (1957) and Atema (1973). With physicist colleagues through the 1980s, Hameroff developed models of microtubules as information processing devices, specifically molecular ('cellular') automata, self-organizing computational devices (Figure 3). Cellular automata are computational systems in which fundamental units, or 'cells' in a grid or lattice can each exist in specific states, e.g. 1 or 0, at a given time (Wolfram, 2002). Each cell interacts with its neighbor cells at discrete, synchronized time steps, the state of each cell at any particular time step determined by its state and its neighbor cell states at the previous time step, and rules governing the interactions. In such ways, using

Cosmology of Consciousness

simple neighbor interactions in simple lattice grids, cellular automata can perform complex computation and generate complex patterns.

Cells in cellular automata are meant to imply fundamental units. But biological cells are not necessarily simple, as illustrated by the clever Paramecium. Molecular automata are cellular automata in which the fundamental units, bits or cells are states of molecules, much smaller than biological cells. A dynamic, interactive molecular grid or lattice is required.

Microtubules are lattices of tubulin dimers which Hameroff and colleagues modeled as molecular automata. Discrete states of tubulin were suggested to act as bits, switching between states, and interacting (via dipole-dipole coupling) with neighbor tubulin bit states in 'molecular automata' computation (Hameroff and Watt, 1982; Rasmussen et al., 1990; Tuszynski et al., 1995). The mechanism for bit-like switching at the level of each tubulin was proposed to depend on the van der Waals-London force in non-polar, water-excluding regions ('hydrophobic pockets') within each tubulin.

Proteins are largely heterogeneous arrays of amino acid residues, including both water-soluble polar and water-insoluble non-polar groups, the latter including phenylalanine and tryptophan with electron resonance clouds (e.g. phenyl and indole rings). Such non-polar groups coalesce during protein folding to form homogeneous water-excluding 'hydrophobic' pockets within which instantaneous dipole couplings between nearby electron clouds operate. These are London forces which are extremely weak but numerous and able to act collectively in hydrophobic regions to influence and determine protein state (Voet and Voet, 1995).

London forces in hydrophobic pockets of various neuronal proteins are the mechanisms by which anesthetic gases selectively erase consciousness (Franks and Lieb, 1984). Anesthetics bind by their own London force attractions with electron clouds of the hydrophobic pocket, presumably impairing normally-occurring London forces governing protein switching required for consciousness (Hameroff, 2006).

In Figure 2, and as previously used in Orch OR, London forces are illustrated in cartoon fashion. A single hydrophobic pocket is depicted in tubulin, with portions of two electron resonance rings in the pocket. Single electrons in each ring repel each other, as their electron cloud net dipole flips (London force oscillation). London forces in hydrophobic pockets were used as the switching mechanism to distinguish discrete states for each tubulin in microtubule automata. In recent years tubulin hydrophobic regions and switching in the Orch OR proposal that we describe below have been clarified and updated (see Section 8).

To synchronize discrete time steps in microtubule automata, tubulins in microtubules were assumed to oscillate synchronously in a manner proposed by Fröhlich for biological coherence. Biophysicist Herbert Fröhlich (1968; 1970; 1975) had suggested that biomolecular dipoles constrained in a common geometry and voltage field would oscillate coherently, coupling, or condensing to a common vibrational mode. He proposed that biomolecular dipole lattices could convert ambient energy to coherent, synchronized dipole excitations, e.g. in the gigahertz (109 s−1) frequency range. Fröhlich coherence or condensation can be either quantum coherence (e.g. Bose-Einstein condensation) or classical synchrony (Reimers et al., 2009).

In recent years coherent excitations have been found in living cells emanating from microtubules at 8 megahertz (Pokorny et al., 2001; 2004). Bandyopadhyay (2011) has found a series of coherence resonance peaks in single microtubules ranging from 12 kilohertz to 8 megahertz.

Rasmussen et al (1990) applied Fröhlich synchrony (in classical mode) as a clocking mechanism for computational time steps in simulated microtubule automata. Based on dipole couplings between neighboring tubulins in the microtubule lattice geometry, they found traveling gliders, complex patterns, computation and learning. Microtubule automata within brain neurons could potentially provide another level of information processing in the brain.

Approximately 10^8 tubulins in each neuron switching and oscillating in the range of 10^7 per second (e.g. Pokorny 8 MHz) gives an information capacity at the microtubule level of 10^{15} operations per second per neuron. This predicted capacity challenged and annoyed AI whose estimates for information processing at the level of neurons and synapses were virtually the same as this single-cell value, but for the entire brain (10^{11} neurons, 10^3 synapses per neuron, 10^2 transmissions per synapse per second = 10^{16} operations per second). Total brain capacity when taken at the microtubule level (in 10^{11} neurons) would potentially be 10^{26} operations per second, pushing the goalpost for AI brain equivalence farther into the future, and down into the quantum regime.

High capacity microtubule-based computing inside brain neurons could account for organization of synaptic regulation, learning and memory, and perhaps act as the substrate for consciousness. But increased brain information capacity per se didn't address most unanswered questions about consciousness (Section 2). Something was missing.

Cosmology of Consciousness

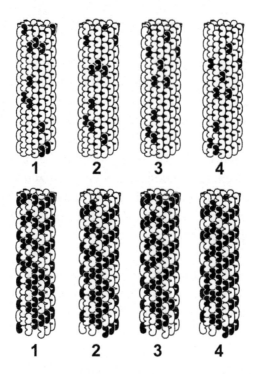

Figure 3. Microtubule automata (Rasmussen et al, 1990). Top: 4 time steps (e.g. at 8 megahertz, Pokorny et al, 2001) showing propagation of information states and patterns ('gliders' in cellular automata parlance). Bottom: At different dipole coupling parameter, bi-directional pattern movement and computation occur.

4. Objective Reduction (OR)

In 1989 Penrose published The Emperor's New Mind, which was followed in 1994 by Shadows of the Mind. Critical of AI, both books argued, by appealing to Gödel's theorem and other considerations, that certain aspects of human consciousness, such as understanding, must be beyond the scope of any computational system, i.e. 'non-computable'. Non-computability is a perfectly well-defined mathematical concept, but it had not previously been considered as a serious possibility for the result of physical actions. The non-computable ingredient required for human consciousness and understanding, Penrose suggested, would have to lie in an area where our current physical theories are fundamentally incomplete, though of important relevance to the scales that are pertinent to the operation of our brains. The only serious possibility was the incompleteness of quantum theory "an incompleteness that both Einstein and Schrödinger had recognized, despite quantum theory having frequently been argued to represent the pinnacle of 20th century scientific achievement. This incompleteness is the unresolved issue referred to as the 'measurement problem', which we consider in more detail below, in Section 5. One way to resolve it would be to provide an extension of the standard framework of quantum mechanics by introducing an objective form of quantum state reduction" termed 'OR' (objective reduction), an idea which we also describe more fully below, in Section 6.

In Penrose (1989), the tentatively suggested OR proposal would have its onset determined by a condition referred to there as 'the one-graviton' criterion. However, in Penrose (1995), a much better-founded criterion was used, now sometimes referred to as the Diósi-Penrose proposal (henceforth 'DP'; see Diósi 1987, 1989, Penrose 1993, 1996, 2000, 2009). This is an objective physical threshold, providing a plausible lifetime for quantum-superposed states. Other such OR proposals had also been put forward, from time to time (e.g. Kibble 1981, Pearle 1989, Pearle and Squires 1994, Ghirardi et al., 1986, 1990; see Ghirardi 2011, this volume) as solutions to the measurement problem, but had not originally been suggested as having anything to do with the consciousness issue. The Diósi-Penrose proposal is sometimes referred to as a 'quantum-gravity' scheme, but it is not part of the normal ideas used in quantum gravity, as will be explained below (Section 6). Moreover, the proposed connection between consciousness and quantum measurement is almost opposite, in the Orch OR scheme, to the kind of idea that had frequently been put forward in the early days of quantum mechanics (see, for example, Wigner 1961) which suggests that a 'quantum measurement' is something that occurs only as a result of the conscious intervention of an observer. This issue, also, will be discussed below (Section 5).

5. The Nature of Quantum Mechanics and its Fundamental Problem

The term 'quantum' refers to a discrete element of energy in a system, such as the energy E of a particle, or of some other subsystem, this energy being related to a fundamental frequency ν of its oscillation, according to Max Planck's famous formula (where h is Planck's constant):

$$E = h\,\nu.$$

This deep relation between discrete energy levels and frequencies of oscillation underlies the wave/particle duality inherent in quantum phenomena. Neither the word "particle" nor the word "wave" adequately conveys the true nature of a basic quantum entity, but both provide useful partial pictures.

The laws governing these submicroscopic quantum entities differ from those governing our everyday classical world. For example, quantum particles can exist in two or more states or locations simultaneously, where such a multiple coexisting superposition of alternatives (each alternative being weighted by a

Cosmology of Consciousness

complex number) would be described mathematically by a quantum wavefunction. We don't see superpositions in the consciously perceived world; we see objects and particles as material, classical things in specific locations and states.

Another quantum property is 'non-local entanglement,' in which separated components of a system become unified, the entire collection of components being governed by one common quantum wavefunction. The parts remain somehow connected, even when spatially separated by significant distances (e.g. over 10 kilometres, Tittel et al., 1998). Quantum superpositions of bit states (quantum bits, or qubits) can be interconnected with one another through entanglement in quantum computers. However, quantum entanglements cannot, by themselves, be used to send a message from one part of an entangled system to another; yet entanglement can be used in conjunction with classical signaling to achieve strange effects--such as the strange phenomenon referred to as quantum teleportation--that classical signalling cannot achieve by itself (e.g. Bennett and Wiesner, 1992; Bennett et al., 1993; Bouwmeester et al., 1997; Macikic et al., 2002).

The issue of why we don't directly perceive quantum superpositions is a manifestation of the measurement problem referred to in Section 4. Put more precisely, the measurement problem is the conflict between the two fundamental procedures of quantum mechanics. One of these procedures, referred to as unitary evolution, denoted here by U, is the continuous deterministic evolution of the quantum state (i.e. of the wavefunction of the entire system) according to the fundamental Schrödinger equation, The other is the procedure that is adopted whenever a measurement of the system--or observation--is deemed to have taken place, where the quantum state is discontinuously and probabilistically replaced by another quantum state (referred to, technically, as an eigenstate of a mathematical operator that is taken to describe the measurement). This discontinuous jumping of the state is referred to as the reduction of the state (or the 'collapse of the wavefunction'), and will be denoted here by the letter R. The conflict that is termed the measurement problem (or perhaps more accurately as the measurement paradox) arises when we consider the measuring apparatus itself as a quantum entity, which is part of the entire quantum system consisting of the original system under observation together with this measuring apparatus. The apparatus is, after all, constructed out of the same type of quantum ingredients (electrons, photons, protons, neutrons etc.--or quarks and gluons etc.) as is the system under observation, so it ought to be subject also to the same quantum laws, these being described in terms of the continuous and deterministic U. How, then, can the discontinuous and probabilistic R come about as a result of the interaction (measurement) between two parts of the quantum system? This is the measurement problem (or paradox).

There are many ways that quantum physicists have attempted to come to terms with this conflict (see, for example, Bell 1966, Bohm 1951, Rae 1994, Polkinghorne 2002, Penrose, 2004). In the early 20th century, the Danish physicist Niels Bohr, together with Werner Heisenberg, proposed the pragmatic 'Copenhagen interpretation', according to which the wavefunction of a quantum system, evolving according to U, is not assigned any actual physical 'reality', but is taken as basically providing the needed 'book-keeping' so that eventually probability values can be assigned to the various possible outcomes of a quantum measurement. The measuring device itself is explicitly taken to behave classically and no account is taken of the fact that the device is ultimately built from quantum-level constituents. The probabilities are calculated, once the nature of the measuring device is known, from the state that the wavefunction has U-evolved to at the time of the measurement. The discontinuous "jump" that the wavefunction makes upon measurement, according to R, is attributed to the change in 'knowledge' that the result of the measurement has on the observer. Since the wavefunction is not assigned physical reality, but is considered to refer merely to the observer's knowledge of the quantum system, the jumping is considered simply to reflect the jump in the observer's knowledge state, rather than in the quantum system under consideration.

Many physicists remain unhappy with such a point of view, however, and regard it largely as a 'stop-gap', in order that progress can be made in applying the quantum formalism, without this progress being held up by a lack of a serious quantum ontology, which might provide a more complete picture of what is actually going on. One may ask, in particular, what it is about a measuring device that allows one to ignore the fact that it is itself made from quantum constituents and is permitted to be treated entirely classically. A good many proponents of the Copenhagen standpoint would take the view that while the physical measuring apparatus ought actually to be treated as a quantum system, and therefore part of an over-riding wavefunction evolving according to U, it would be the conscious observer, examining the readings on that device, who actually reduces the state, according to R, thereby assigning a physical reality to the particular observed alternative resulting from the measurement. Accordingly, before the intervention of the observer's consciousness, the various alternatives of the result of the measurement including the different states of the measuring apparatus would, in effect, still coexist in superposition, in accordance with what would be the usual evolution according to U. In this way, the Copenhagen viewpoint puts consciousness outside science, and does not seriously address the nature and physical role of superposition itself nor the question of how large quantum superpositions like Schrödinger's superposed live and dead cat (see below) might actually become one thing or another.

A more extreme variant of this approach is the 'multiple worlds hypothesis' of Everett (1957) in which each possibility in a superposition evolves to form its own universe, resulting in an infinite multitude of coexisting 'parallel' worlds. The stream of consciousness of the observer is supposed somehow to 'split', so that there is one in each of the worlds--at least in those worlds for which the observer remains alive and conscious. Each instance of the observer's consciousness experiences a separate independent world, and is not directly aware of any of the other worlds.

A more 'down-to-earth' viewpoint is that of environmental decoherence, in which interaction of a superposition with its environment 'erodes' quantum states, so that instead of a single wavefunction being used to describe the state, a more complicated entity is used, referred to as a density matrix. However decoherence does not provide a consistent ontology for the reality of the world, in relation to the density matrix (see, for example, Penrose 2004, Sections 29.3-6), and provides merely a pragmatic procedure. Moreover, it does not address the issue of how R might arise in isolated systems, nor the nature of isolation, in which an external 'environment' would not be involved, nor does it tell us which part of a system is to be regarded as the 'environment' part, and it provides no limit to the size of that part which can remain subject to quantum superposition.

Still other approaches include various types of objective reduction (OR) in which a specific objective threshold is proposed to cause quantum state reduction (e.g. Kibble 1981; Pearle 1989; Ghirardi et al., 1986; Percival, 1994; Ghirardi, 2011). The specific OR scheme that is used in Orch OR will be described in Section 6.

The quantum pioneer Erwin Schrödinger took pains to point out the difficulties that confront the U-evolution of a quantum system with his still-famous thought experiment called 'Schrödinger's cat'. Here, the fate of a cat in a box is determined by magnifying a quantum event (say the decay of a radioactive atom, within a specific time period that would provide a 50% probability of decay) to a macroscopic action which would kill the cat, so that according to Schrödinger's own U-evolution the cat would be in a quantum superposition of being both dead and alive at the same time. If this U-evolution is maintained until the box is opened and the cat observed, then it would have to be the conscious human observing the cat that results in the cat becoming either dead or alive (unless, of course, the cat's own consciousness could be considered to have already served this purpose). Schrödinger intended to illustrate the absurdity of the direct applicability of the rules of quantum mechanics (including his own U-evolution) when applied at the level of a cat. Like Einstein, he regarded quantum mechanics as an incomplete theory, and his 'cat' provided an excellent example for emphasizing this incompleteness. There is a need for something to be done about quantum mechanics, irrespective of the issue of its relevance to consciousness.

6. The Orch OR Scheme

Orch OR depends, indeed, upon a particular OR extension of current quantum mechanics, taking the bridge between quantum- and classical-level physics as a 'quantum-gravitational' phenomenon. This is in contrast with the various conventional viewpoints (see Section 5), whereby this bridge is claimed to result, somehow, from 'environmental decoherence', or from 'observation by a conscious observer', or from a 'choice between alternative worlds', or some other interpretation of how the classical world of one actual alternative may be taken to arise out of fundamentally quantum-superposed ingredients.

It must also be made clear that the Orch OR scheme involves a different interpretation of the term 'quantum gravity' from what is usual. Current ideas of quantum gravity (see, for example Smolin, 2002) normally refer, instead, to some sort of physical scheme that is to be formulated within the bounds of standard quantum field theory--although no particular such theory, among the multitude that has so far been put forward, has gained anything approaching universal acceptance, nor has any of them found a fully consistent, satisfactory formulation. 'OR' here refers to the alternative viewpoint that standard quantum (field) theory is not the final answer, and that the reduction R of the quantum state ('collapse of the wavefunction') that is adopted in standard quantum mechanics is an actual physical phenomenon which is not part of the conventional unitary formalism U of quantum theory (or quantum field theory) and does not arise as some kind of convenience or effective consequence of environmental decoherence, etc., as the conventional U formalism would seem to demand. Instead, OR is taken to be one of the consequences of melding together the principles of Einstein's general relativity with those of the conventional unitary quantum formalism U, and this demands a departure from the strict rules of U. According to this OR viewpoint, any quantum measurement--whereby the quantum-superposed alternatives produced in accordance with the U formalism becomes reduced to a single actual occurrence--is real objective physical phenomenon, and it is taken to result from the mass displacement between the alternatives being sufficient, in gravitational terms, for the superposition to become unstable.

In the DP (Diósi-Penrose) scheme for OR, the superposition reduces to one of the alternatives in a time scale τ that can be estimated (for a superposition of two states each of which can be taken to be stationary on its own) according to the formula

$$\tau \approx \hbar/E_G.$$

Here $\hbar$ (=$h/2\pi$) is Dirac's form of Planck's constant h and E_G is the gravitational self-energy of the difference between the two mass distributions of the superposition. (For a superposition for which each mass distribution is a rigid translation of the other, E_G is the energy it would cost to displace one component of the superposition in the gravitational field of the other, in moving it from coincidence to the quantum-displaced location; see Diósi 1989, Penrose 1993, 2000, 2009).

According to Orch OR, the (objective) reduction is not the entirely random process of standard theory, but acts according to some non-computational new physics (see Penrose 1989, 1994). The idea is that consciousness is associated with this (gravitational) OR process, but occurs significantly only when the alternatives are part of some highly organized structure, so that such occurrences of OR occur in an extremely orchestrated form. Only then does a recognizably conscious event take place. On the other hand, we may consider that any individual occurrence of OR would be an element of proto-consciousness.

The OR process is considered to occur when quantum superpositions between slightly differing space-times take place, differing from one another by an integrated space-time measure which compares with the fundamental and extremely tiny Planck (4-volume) scale of space-time geometry. Since this is a 4-volume Planck measure, involving both time and space, we find that the time measure would be particularly tiny when the space-difference measure is relatively large (as with Schrödinger's cat), but for extremely tiny space-difference measures, the time measure might be fairly long, such as some significant fraction of a second. We shall be seeing this in more detail shortly, together with its particular relevance to microtubules. In any case, we recognize that the elements of proto-consciousness would be intimately tied in with the most primitive Planck-level ingredients of space-time geometry, these presumed 'ingredients' being taken to be at the absurdly tiny level of 10^{-35}m and 10^{-43}s, a distance and a time some 20 orders of magnitude smaller than those of normal particle-physics scales and their most rapid processes. These scales refer only to the normally extremely tiny differences in space-time geometry between different states in superposition, and OR is deemed to take place when such space-time differences reach the Planck level. Owing to the extreme weakness of gravitational forces as compared with those of the chemical and electric forces of biology, the energy E_G is liable to be far smaller than any energy that arises directly from biological processes.

However, E_G is not to be thought of as being in direct competition with any of the usual biological energies, as it plays a completely different role, supplying a needed energy uncertainty that then allows a choice to be made between the separated space-time geometries. It is the key ingredient of the computation of the reduction time τ. Nevertheless, the extreme weakness of gravity tells us there must be a considerable amount of material involved in the coherent mass displacement between superposed structures in order that τ can be small enough to be playing its necessary role in the relevant OR processes in the brain. These superposed structures should also process information and regulate neuronal physiology. According to Orch OR, microtubules are central to these structures, and some form of biological quantum computation in microtubules (most probably primarily in the more symmetrical A-lattice microtubules) would have to have evolved to provide a subtle yet direct connection to Planck-scale geometry, leading eventually to discrete moments of actual conscious experience.

The degree of separation between the space-time sheets is mathematically described in terms of a symplectic measure on the space of 4-dimensional metrics (cf. Penrose, 1993). The separation is, as already noted above, a space-time separation, not just a spatial one. Thus the time of separation contributes as well as the spatial displacement. Roughly speaking, it is the product of the temporal separation T with the spatial separation S that measures the overall degree of separation, and OR takes place when this overall separation reaches a critical amount. This critical amount would be of the order of unity, in absolute units, for which the Planck-Dirac constant $\hbar$, the gravitational constant G, and the velocity of light c, all take the value unity, cf. Penrose, 1994 - pp. 337-339. For small S, the lifetime $\tau \approx T$ of the superposed state will be large; on the other hand, if S is large, then τ will be small.

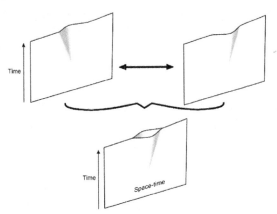

Figure 4. From Penrose, 1994 (P. 338). With four spatiotemporal dimensions condensed to a 2-dimensional spacetime sheet, mass location may be represented as a particular curvature of that sheet, according to general relativity. Top: Two different mass locations as alternative spacetime curvatures. Bottom: a bifurcating spacetime is depicted as the union ("glued together version") of the

Cosmology of Consciousness

two alternative spacetime histories that are depicted at the top of the Figure. Hence a quantum superposition of simultaneous alternative locations may be seen as a separation in fundamental spacetime geometry.

To estimate S, we compute (in the Newtonian limit of weak gravitational fields) the gravitational self-energy E_G of the difference between the mass distributions of the two superposed states. (That is, one mass distribution counts positively and the other, negatively; see Penrose, 1993; 1995.) The quantity S is then given by:

$$S \approx EG$$

and $T \approx \tau$, whence

$$\tau \approx \hbar/EG \text{ , i.e. } EG \approx \hbar/\tau.$$

Thus, the DP expectation is that OR occurs with the resolving out of one particular space-time geometry from the previous superposition when, on the average, $\tau \approx \hbar/E_G$. Moreover, according to Orch OR, this is accompanied by an element of proto-consciousness.

Environmental decoherence need play no role in state reduction, according to this scheme. The proposal is that state reduction simply takes place spontaneously, according to this criterion. On the other hand, in many actual physical situations, there would be much material from the environment that would be entangled with the quantum-superposed state, and it could well be that the major mass displacement--and therefore the major contribution to E_G --would occur in the environment rather than in the system under consideration. Since the environment will be quantum-entangled with the system, the state-reduction in the environment will effect a simultaneous reduction in the system. This could shorten the time for the state reduction R to take place very considerably. It would also introduce an uncontrollable random element into the result of the reduction, so that any non-random (albeit non-computable, according to Orch OR) element influencing the particular choice of state that is actually resolved out from the superposition would be completely masked by this randomness. In these circumstances the OR-process would be indistinguishable from the R-process of conventional quantum mechanics. If the suggested non-computable effects of this OR proposal are to be laid bare, if E_G is to be able to evolve and be orchestrated for conscious moments, we indeed need significant isolation from the environment.

As yet, no experiment has been refined enough to determine whether this (DP) OR proposal is actually respected by Nature, but the experimental testing of the scheme is fairly close to the borderline of what can be achieved with present-day technology (see, for example, Marshall et al. 2003). One ought to begin to see the effects of this OR scheme if a small object, such as a 10-micron cube of crystalline material could be held in a superposition of two locations, differing by about the diameter of an atomic nucleus, for some seconds, or perhaps minutes.

A point of importance, in such proposed experiments, is that in order to calculate E_G it may not be enough to base the calculation on an average density of the material in the superposition, since the mass will be concentrated in the atomic nuclei, and for a displacement of the order of the diameter of a nucleus, this inhomogeneity in the density of the material can be crucial, and can provide a much larger value for E_G than would be obtained if the material is assumed to be homogeneous. The Schrödinger equation (more correctly, in the zero-temperature approximation, the Schrödinger--Newton equation, see Penrose 2000; Moroz et al. 1998) for the static unsuperposed material would have to be solved, at least approximately, in order to derive the expectation value of the mass distribution, where there would be some quantum spread in the locations of the particles constituting the nuclei.

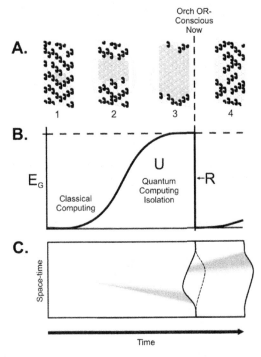

Figure 5. Three descriptions of an Orch OR conscious event by $E_G = \hbar/\tau$. A. Microtubule automata. Quantum (gray) tubulins evolve to meet threshold after Step 3, a moment of consciousness occurs and tubulin states are selected. For actual event (e.g. 25 msec), billions of tubulins are required; a small number is used here for illustration. B. Schematic showing U-like evolution until threshold. C. Space-time sheet with superposition separation reaches threshold and selects one reality/spacetime curvature.

Cosmology of Consciousness

For Orch OR to be operative in the brain, we would need coherent superpositions of sufficient amounts of material, undisturbed by environmental entanglement, where this reduces in accordance with the above OR scheme in a rough time scale of the general order of time for a conscious experience to take place. For an ordinary type of experience, this might be say about $\tau = 10-1s$ which concurs with neural correlates of consciousness, such as particular frequencies of electroencephalograhy (EEG).

Penrose (1989; 1994) suggested that processes of the general nature of quantum computations were occurring in the brain, terminated by OR. In quantum computers (Benioff 1982, Deutsch 1985, Feynman 1986), information is represented not just as bits of either 1 or 0, but also as quantum superposition of both 1 and 0 together (quantum bits or qubits) where, moreover, large-scale entanglements between qubits would also be involved. These qubits interact and compute following the Schrödinger equation, potentially enabling complex and highly efficient parallel processing. As envisioned in technological quantum computers, at some point a measurement is made causing quantum state reduction (with some randomness introduced). The qubits reduce, or collapse to classical bits and definite states as the output.

The proposal that some form of quantum computing could be acting in the brain, this proceeding by the Schrödinger equation without decoherence until some threshold for self-collapse due to a form of non-computable OR could be reached, was made in Penrose 1989. However, no plausible biological candidate for quantum computing in the brain had been available to him, as he was then unfamiliar with microtubules.

7. Penrose-Hameroff Orchestrated Objective Reduction ('Orch OR')

Penrose and Hameroff teamed up in the early 1990s. Fortunately, by then, the DP form of OR mechanism was at hand to be applied to the microtubule-automata models for consciousness as developed by Hameroff. A number of questions were addressed.

How does $\tau \approx \hbar/E_G$ relate to consciousness? Orch OR considers consciousness as a sequence of discrete OR events in concert with neuronal-level activities. In $\tau \approx \hbar/E_G$, τ is taken to be the time for evolution of the pre-conscious quantum wavefunction between OR events, i.e. the time interval between conscious moments, during which quantum superpositions of microtubule states evolve according to the continuous Schrödinger equation before reaching (on the average) the $\tau \approx \hbar/E_G$ OR threshold in time τ, when quantum state reduction and a moment of conscious awareness occurs (Figure 5).

The best known temporal correlate for consciousness is gamma synchrony EEG, 30 to 90 Hz, often referred to as coherent 40 Hz. One possible viewpoint might be to take this oscillation to represent a succession of 40 or so conscious moments per second (τ=25 milliseconds). This would be reasonably consistent with neuroscience (gamma synchrony), with certain ideas expressed in philosophy (e.g. Whitehead 'occasions of experience'), and perhaps even with ancient Buddhist texts which portray consciousness as 'momentary collections of mental phenomena' or as 'distinct, unconnected and impermanent moments which perish as soon as they arise.' (Some Buddhist writings quantify the frequency of conscious moments. For example the Sarvaastivaadins, according to von Rospatt 1995, described 6,480,000 'moments' in 24 hours--an average of one 'moment' per 13.3 msec, ~75 Hz--and some Chinese Buddhism as one "thought" per 20 msec, i.e. 50 Hz.) These accounts, even including variations in frequency, could be considered to be consistent with Orch OR events in the gamma synchrony range. Accordingly, on this view, gamma synchrony, Buddhist 'moments of experience', Whitehead 'occasions of experience', and our proposed Orch OR events might be viewed as corresponding tolerably well with one another.

Putting τ=25msec in EG $\approx\hbar/\tau$, we may ask what is EG in terms of superpositioned microtubule tubulins? EG may be derived from details about the superposition separation of mass distribution. Three types of mass separation were considered in Hameroff-Penrose 1996a for peanut-shaped tubulin proteins of 110,000 atomic mass units: separation at the level of (1) protein spheres, e.g. by 10 percent volume, (2) atomic nuclei (e.g. carbon, ~ 2.5 Fermi length), (3) nucleons (protons and neutrons). The most plausible calculated effect might be separation at the level of atomic nuclei, giving EG as superposition of 2 x 1010 tubulins reaching OR threshold at 25 milliseconds.

Brain neurons each contain roughly 108 tubulins, so only a few hundred neurons would be required for a 25msec, gamma synchrony OR event if 100 percent of tubulins in those neurons were in superposition and avoided decoherence. It seems more likely that a fraction of tubulins per neuron are in superposition. Global macroscopic states such as superconductivity ensue from quantum coherence among only very small fractions of components. If 1 percent of tubulins within a given set of neurons were coherent for 25msec, then 20,000 such neurons would be required to elicit OR. In human brain, cognition and consciousness are, at any one time, thought to involve tens of thousands of neurons. Hebb's (1949) 'cell assemblies', Eccles's (1992) 'modules', and Crick and Koch's (1990) 'coherent sets of neurons' are each estimated to contain some 10,000 to 100,000 neurons which may be widely distributed throughout the brain (Scott, 1995).

Adopting $\tau\approx\hbar/E_G$, we find that, with this point of view with regard to Orch-OR, a spectrum of possible types of conscious event might be able to occur, including those at higher frequency and intensity. It may be noted that Tibetan monk meditators have been found to have 80 Hz gamma synchrony, and perhaps more intense experience (Lutz et al. 2004). Thus, according to the viewpoint proposed above, where we interpret this frequency to be associated with a succession of Orch-OR moments, then $E_G \approx \hbar/\tau$ would appear to require that there is twice as much brain involvement required for 80 Hz than for consciousness occurring at 40 Hz (or $\sqrt{2}$ times as much if the displacement is entirely coherent, since then the mass enters quadratically in E_G). Even higher (frequency), expanded awareness states of consciousness might be expected, with more neuronal brain involvement.

On the other hand, we might take an alternative viewpoint with regard to the probable frequency of Orch-OR actions, and to the resulting frequency of elements of conscious experience. There is the possibility that the discernable moments of consciousness are events that normally occur at a much slower pace than is suggested by the considerations above, and that they happen only at rough intervals of the order of, say, one half a second or so, i.e. ~500msec, rather than ~25msec. One might indeed think of conscious influences as perhaps being rather slow, in contrast with the great deal of vastly faster unconscious computing that might be some form of quantum computing, but without OR. At the present stage of uncertainty about such matters it is perhaps best not to be dogmatic about how the ideas of Orch OR are to be applied. In any case, the numerical assignments provided above must be considered to be extremely rough, and at the moment we are far from being in a position to be definitive about the precise way in which the Orch-OR is to operate. Alternative possibilities will need to be considered with an open mind.

How do microtubule quantum computation avoid decoherence? Technological quantum computers using e.g. ion traps as qubits are plagued by decoherence, disruption of delicate quantum states by thermal vibration, and require extremely cold temperatures and vacuum to operate. Decoherence must be avoided during the evolution toward time τ ($\approx\hbar/E_G$), so that the non-random (non-computable) aspects of OR can be playing their roles. How does quantum computing avoid decoherence in the 'warm, wet and noisy' brain?

It was suggested (Hameroff and Penrose, 1996a) that microtubule quantum states avoid decoherence by being pumped, laser-like, by Fröhlich resonance, and shielded by ordered water, C-termini Debye layers, actin gel and strong mitochondrial electric fields. Moreover quantum states in Orch OR are proposed to originate in hydrophobic pockets in tubulin interiors, isolated from polar interactions, and involve superposition of only atomic nuclei separation.

Moreover, geometrical resonances in microtubules, e.g. following helical pathways of Fibonacci geometry are suggested to enable topological quantum computing and error correction, avoiding decoherence perhaps effectively indefinitely (Hameroff et al 2002) as in a superconductor.

The analogy with high-temperature superconductors may indeed be appropriate, in fact. As yet, there is no fully accepted theory of how such superconductors operate, avoiding loss of quantum coherence from the usual processes of environmental decoherence. Yet there are materials which support superconductivity at temperatures roughly halfway between room temperature and absolute zero (He et al., 2010). This is still a long way from body temperature, of course, but there is now some experimental evidence (Bandyopadhyay 2011) that is indicative of something resembling superconductivity (referred to as 'ballistic conductance'), that occurs in living A-lattice microtubules at body temperature. This will be discussed below.

Physicist Max Tegmark (2000) published a critique of Orch OR based on his calculated decoherence times for microtubules of 10^{-13} seconds at biological temperature, far too brief for physiological effects. However Tegmark didn't include Orch OR stipulations and in essence created, and then refuted his own quantum microtubule model. He assumed superpositions of solitons separated from themselves by a distance of 24 nanometers along the length of the microtubule. As previously described, superposition separation in Orch OR is at the Fermi length level of atomic nuclei, i.e. 7 orders of magnitude smaller than Tegmark's separation value, thus underestimating decoherence time by 7 orders of magnitude, i.e. from 10^{-13} secs to microseconds at 10^{-6} seconds. Hagan et al (2001) used Tegmark's same formula and recalculated microtubule decoherence times using Orch OR stipulations, finding 10^{-4} to 10^{-3} seconds, or longer due to topological quantum effects. It seemed likely biology had evolved optimal information processing systems which can utilize quantum computing, but there was no real evidence either way.

Beginning in 2003, published research began to demonstrate quantum coherence in warm biological systems. Ouyang and Awschalom (2003) showed that quantum spin transfer through phenyl rings (the same as those in protein hydrophobic pockets) is enhanced at increasingly warm temperatures. Other studies showed that quantum coherence occurred at ambient temperatures in proteins involved in photosynthesis, that plants routinely use quantum coherence to produce chemical energy and food (Engel et al, 2007). Further research has demonstrated warm quantum effects in bird brain navigation (Gauger et al, 2011), ion channels (Bernroider and Roy, 2005), sense of smell (Turin, 1996), DNA (Rieper et al., 2011), protein folding (Luo and Lu, 2011), biological water (Reiter et al., 2011) and microtubules.

Recently Anirban Bandyopadhyay and colleagues at the National Institute of Material Sciences in Tsukuba, Japan have used nanotechnology to study electronic conductance properties of single microtubules assembled from porcine brain tubulin. Their preliminary findings (Bandyopadhyay, 2011) include: (1) Microtubules have 8 resonance peaks for AC stimulation (kilohertz to 10 megahertz) which appear to correlate with various helical conductance pathways around the geometric microtubule lattice. (2) Excitation at these resonant frequencies causes microtubules to assemble extremely rapidly, possibly due to Fröhlich condensation. (3) In assembled microtubules AC excitation at resonant frequencies causes electronic conductance to become lossless, or 'ballistic', essentially quantum conductance, presumably along these helical quantum channels. Resonance in the range of kilohertz demonstrates microtubule decoherence times of at least 0.1 millisecond. (4) Eight distinct quantum interference patterns from a single microtubule, each correlating with one of the 8 resonance frequencies and pathways. (5) Ferroelectric hysteresis demonstrates memory capacity in microtubules. (6) Temperature-independent conductance also suggests quantum effects. If confirmed, such findings would demonstrate Orch OR to be biologically feasible.

How does microtubule quantum computation and Orch OR fit with recognized neurophysiology? Neurons are composed of multiple dendrites and a cell body/ soma which receive and integrate synaptic inputs to a threshold for firing outputs along a single axon. Microtubule quantum computation in Orch OR is assumed to occur in dendrites and cell bodies/soma of brain neurons, i.e. in regions of integration of inputs in integrate-and-fire neurons. As opposed to axonal firings, dendritic/somatic integration correlates best with local field potentials, gamma synchrony EEG, and action of anesthetics erasing consciousness. Tononi (2004) has identified integration of information as the neuronal function most closely associated with consciousness. Dendritic microtubules are uniquely arranged in local mixed polarity networks, well-suited for integration of synaptic inputs.

Membrane synaptic inputs interact with post-synaptic microtubules by activation of microtubule-associated protein 2 ('MAP2', associated with learning), and calcium-calmodulin kinase II (CaMKII, Hameroff et al, 2010). Such inputs were suggested by Penrose and Hameroff (1996a) to 'tune', or 'orchestrate' OR-mediated quantum computations in microtubules by MAPs, hence 'orchestrated objective reduction', 'Orch OR'.

Proposed mechanisms for microtubule avoidance of decoherence were described above, but another question remains. How would microtubule quantum computations which are isolated from the environment, still interact with that environment for input and output? One possibility that Orch OR suggests is that perhaps phases of isolated quantum computing alternate with phases of classical environmental interaction, e.g. at gamma synchrony, roughly 40 times per second. (Computing pioneer Paul Benioff suggested such a scheme of alternating quantum and classical phases in a science fiction story about quantum computing robots.)

With regard to outputs resulting from processes taking place at the level of microtubules in Orch-OR quantum computations, dendritic/somatic microtubules receive and integrate synaptic inputs during classical phase. They then become isolated quantum computers and evolve to threshold for Orch OR at which they reduce their quantum states at an average time interval τ (given by by $\tau \approx \hbar/EG$). The particular tubulin states chosen in the reduction can then trigger axonal firing, adjust firing threshold, regulate synapses and encode memory. Thus Orch OR can have causal efficacy in conscious actions and behavior, as well as providing conscious experience and memory.

Orch OR in evolution In the absence of Orch OR, non-conscious neuronal activities might proceed by classical neuronal and microtubule-based computation. In addition there could be quantum computations in microtubules that do not reach the Orch OR level, and thereby also remain unconscious.

This last possibility is strongly suggested by considerations of natural selection, since some relatively primitive microtubule infrastructure, still able to support quantum computation, would have to have preceded the more sophisticated kind that we now find in conscious animals. Natural selection proceeds in steps, after all, and one would not expect that the capability of the substantial level of coherence across the brain that would be needed for the non-computable OR of human conscious understanding to be reached, without something more primitive having preceded it. Microtubule quantum computing by U evolution which avoids decoherence would well be advantageous to biological processes without ever reaching threshold for OR.

Microtubules may have appeared in eukaryotic cells 1.3 billion years ago due to symbiosis among prokaryotes, mitochondria and spirochetes, the latter the apparent origin of microtubules which provided movement to previously immobile cells (e.g. Margulis and Sagan, 1995). Because Orch OR depends on $\tau \approx \hbar/E_G$, more primitive consciousness in simple, small organisms would involve smaller E_G, and longer times τ to avoid decoherence. As simple nervous systems and arrangements of microtubules grew larger and developed anti-decoherence mechanisms, inevitably a system would avoid decoherence long enough to reach threshold for Orch OR conscious moments. Central nervous systems around 300 neurons, such as those present at the early Cambrian evolutionary explosion 540 million years ago, could have τ near one minute, and thus be feasible in terms of avoiding decoherence (Hameroff, 1998d). Perhaps the onset of Orch OR and consciousness with relatively slow and simple conscious moments, precipitated the accelerated evolution.

Only at a much later evolutionary stage would the selective advantages of a capability for genuine understanding come about. This would require the non-computable capabilities of Orch OR that go beyond those of mere quantum computation, and depend upon larger scale infrastructure of efficiently functioning microtubules, capable of operating quantum-computational processes. Further evolution providing larger sets of microtubules (larger E_G) able to be isolated from decoherence would enable, by $\tau \approx \hbar/E_G$, more frequent and more intense moments of conscious experience. It appears human brains could have evolved to having Orch OR conscious moments perhaps as frequently as every few milliseconds.

How could microtubule quantum states in one neuron extend to those in other neurons throughout the brain? Assuming microtubule quantum state phases are isolated in a specific neuron, how could that quantum state involve microtubules in other neurons throughout the brain without traversing membranes and synapses? Orch OR proposes that quantum states can extend by tunneling, leading to entanglement between adjacent neurons through gap junctions.

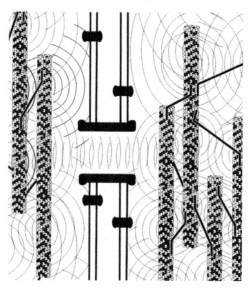

Figure 6. Portions of two neurons connected by a gap junction with microtubules (linked by microtubule-associated proteins, 'MAPs') computing via states (here represented as black or white) of tubulin protein subunits. Wavy lines suggest entanglement among quantum states (not shown) in microtubules.

Gap junctions are primitive electrical connections between cells, synchronizing electrical activities. Structurally, gap junctions are windows between cells which may be open or closed. When open, gap junctions synchronize adjacent cell membrane polarization states, but also allow passage of molecules between cytoplasmic compartments of the two cells. So both membranes and cytoplasmic interiors of gap-junction-connected neurons are continuous, essentially one complex 'hyper-neuron' or syncytium. (Ironically, before Ramon-y-Cajal showed that neurons were discrete cells, the prevalent model for brain structure was a continuous threaded-together syncytium as proposed by Camille Golgi.) Orch OR suggests that quantum states in microtubules in one neuron could extend by entanglement and tunneling through gap junctions to microtubules in adjacent neurons and glia (Figure 6), and from those cells to others, potentially in brain-wide syncytia.

Open gap junctions were thus predicted to play an essential role in the neural correlate of consciousness (Hameroff, 1998a). Beginning in 1998, evidence began to show that gamma synchrony, the best measureable correlate of consciousness, depended on gap junctions, particularly dendritic-dendritic gap junctions (Dermietzel, 1998; Draguhn et al, 1998; Galaretta and Hestrin, 1999). To account for the distinction between conscious activities and non-conscious 'auto-pilot' activities, and the fact that consciousness can occur in various brain regions, Hameroff (2009) developed the "Conscious pilot" model in which syncytial zones of dendritic gamma synchrony move around the brain, regulated by gap junction openings and closings, in turn regulated by microtubules. The model suggests consciousness literally moves around the brain in a mobile synchronized zone, within which isolated, entangled microtubules carry out quantum computations and Orch OR. Taken together, Orch OR and the conscious pilot distinguish conscious from non-conscious functional processes in the brain.

Libet's backward time referral In the 1970s neurophysiologist Benjamin Libet performed experiments on patients having brain surgery while awake, i.e. under local anesthesia (Libet et al., 1979). Able to stimulate and record from conscious human brain, and gather patients' subjective reports with precise timing, Libet determined that conscious perception of a stimulus required up to 500 msec of brain activity post-stimulus, but that conscious awareness occurred at 30 msec post-stimulus, i.e. that subjective experience was referred 'backward in time'.

Bearing such apparent anomalies in mind, Penrose put forward a tentative suggestion, in The Emperor's New Mind, that effects like Libet's backward time referral might be related to the fact that quantum entanglements are not mediated in a normal causal way, so that it might be possible for conscious experience not to follow the normal rules of sequential time progression, so long as this does not lead to contradictions with external causality. In Section 5, it was pointed out that the (experimentally confirmed) phenomenon of 'quantum teleportation' (Bennett et al., 1993; Bouwmeester et al., 1997; Macikic et al., 2002) cannot be explained in terms of ordinary classical information processing, but as a combination of such classical causal influences and the acausal effects of quantum entanglement. It indeed turns out that quantum entanglement effects--referred to as 'quantum information' or 'quanglement' (Penrose 2002, 2004)--appear to have to be thought of as being able to propagate in either direction in time (into the past or into the future). Such effects, however, cannot by themselves be used to communicate ordinary information into the past. Nevertheless, in conjunction with normal classical future-propagating (i.e. 'causal') signalling, these quantum-teleportation influences can achieve certain kinds of 'signalling' that cannot be achieved simply by classical future-directed means.

The issue is a subtle one, but if conscious experience is indeed rooted in the OR process, where we take OR to relate the classical to the quantum world, then apparent anomalies in the sequential aspects of consciousness are perhaps to be expected. The Orch OR scheme allows conscious experience to be temporally non-local to a degree, where this temporal non-locality would spread to the kind of time scale τ that would be involved in the relevant Orch OR process, which might indeed allow this temporal non-locality to spread to a time $\tau=500$ms. When the 'moment' of an internal conscious experience is timed externally, it may well be found that this external timing does not precisely accord with a time progression that would seem to apply to internal conscious experience, owing to this temporal non-locality intrinsic to Orch OR.

Measurable brain activity correlated with a stimulus often occurs several hundred msec after that stimulus, as Libet showed. Yet in activities ranging from rapid conversation to competitive athletics, we respond to a stimulus (seemingly consciously) before the above activity that would be correlated with that stimulus occurring in the brain. This is interpreted in conventional neuroscience and philosophy (e.g. Dennett, 1991; Wegner, 2002) to imply that in such cases we respond non-consciously, on auto-pilot, and subsequently have only an illusion of conscious response. The mainstream view is that consciousness is epiphenomenal illusion, occurring after-the-fact as a false impression of conscious control of behavior. We are merely 'helpless spectators' (Huxley, 1986).

However, the effective quantum backward time referral inherent in the temporal non-locality resulting from the quanglement aspects of Orch OR, as suggested above, enables conscious experience actually to be temporally non-local, thus providing a means to rescue consciousness from its unfortunate characterization as epiphenomenal illusion. Accordingy, Orch OR could well enable consciousness to have a causal efficacy, despite its apparently anomalous relation to a timing assigned to it in relation to an external clock, thereby allowing conscious action to provide a semblance of free will.

8. Orch OR Criticisms and Responses

Orch OR has been criticized repeatedly since its inception. Here we review and summarize major criticisms and responses.

Grush and Churchland, 1995. Philosophers Grush and Churchland (1995) took issue with the Gödel's theorem argument, as well as several biological factors. One objection involved the microtubule-disabling drug colchicine which treats diseases such as gout by immobilizing neutrophil cells which cause painful inflammation in joints. Neutrophil mobility requires cycles of microtubule assembly/disassembly, and colchicine prevents re-assembly, impairing neutrophil mobility and reducing inflammation. Grush and Churchland pointed out that patients given colchicine do not lose consciousness, concluding that microtubules cannot be essential for consciousness. Penrose and Hameroff (1995) responded point-by-point to every objection, e.g. explaining that colchicine does not cross the blood brain barrier, and so doesn't reach the brain. Colchicine infused directly into the brains of animals does cause severe cognitive impairment and apparent loss of consciousness (Bensimon and Chemat, 1991).

Tuszynski et al, 1998. Tuszynski et al (1998) questioned how extremely weak gravitational energy in Diósi-Penrose OR could influence tubulin protein states. In Hameroff and Penrose (1996a), the gravitational self-energy E_G for tubulin superposition was calculated for separation of tubulin from itself at the level of its atomic nuclei. Because the atomic (e.g. carbon) nucleus displacement is greater than its radius (the nuclei separate completely), the gravitational self-energy E_G is given by: $E_G=Gm^2/ac$, where ac is the carbon nucleus sphere radius equal to 2.5 Fermi distances, m is the mass of tubulin, and G is the gravitational constant. Brown and Tuszynski calculated E_G (using separation at the nanometer level of the entire tubulin protein), finding an appropriately small energy E of 10^{-27} electron volts (eV) per tubulin, infinitesimal compared with ambient energy kT of 10^{-4}eV. Correcting for the smaller superposition separation distance of 2.5 Fermi lengths in Orch OR gives a significantly larger, but still tiny 10^{-21}eV per tubulin. With $2\text{-}10^{10}$ tubulins per 25msec, the conscious Orch OR moment would be roughly 10^{-10}eV (10^{-29} joules), still insignificant compared to kT at 10^{-4}eV.

All this serves to illustrate the fact that the energy EG does not actually play a role in physical processes as an energy, in competition with other energies that are driving the physical (chemical, electronic) processes of relevance. In a clear sense E_G is, instead, an energy uncertainty--and it is this uncertainty that allows quantum state reduction to take place without violation of energy conservation. The fact that E_G is far smaller than the other energies involved in the relevant physical processes is a necessary feature of the consistency of the OR scheme. It does not supply the energy to drive the physical processes involved, but it provides the energy uncertainty that allows the freedom for processes having virtually the same energy as each other to be alternative actions. In practice, all that E_G is needed for is to tell us how to calculate the lifetime τ of the superposition. E_G would enter into issues of energy balance only if gravitational interactions between the parts of the system were important in the processes involved. (The Earth's gravitational field plays no role in this either, because it cancels out in the calculation of E_G.) No other forces of nature directly contribute to E_G, which is just as well, because if they did, there would be a gross discrepancy with observational physics.

Tegmark, 2000. Physicist Max Tegmark (2000) confronted Orch OR on the basis of decoherence. This was discussed at length in Section 7.

Koch and Hepp, 2006. In a challenge to Orch OR, neuroscientists/physicists Koch and Hepp published a thought experiment in Nature, describing a person observing a superposition of a cat both dead and alive with one eye, the other eye distracted by a series of images (binocular rivalry). They asked 'Where in the observer's brain would reduction occur?', apparently assuming Orch OR followed the Copenhagen interpretation in which conscious observation causes quantum state reduction. This is precisely the opposite of Orch OR in which consciousness is the orchestrated quantum state reduction given by OR.

Orch OR can account for the related issue of bistable perceptions (e.g. the famous face/vase illusion, or Necker cube). Non-conscious superpositions of both possibilities (face and vase) during pre-conscious quantum superposition then reduce by OR at time τ to conscious perception of one or the other, face or vase. The reduction would occur among microtubules within neurons interconnected by gap junctions in various areas of visual and pre-frontal cortex and other brain regions.

Reimers et al (2009) described three types of Fröhlich condensation (weak, strong and coherent, the first classical and the latter two quantum). They validated 8 MHz coherence measured in microtubules by Pokorny (2001; 2004) as weak condensation. Based on simulation of a 1-dimensional linear chain of tubulin dimers representing a microtubule, they concluded only weak Fröhlich condensation occurs in microtubules. Claiming Orch OR requires strong or coherent Fröhlich condensation, they concluded Orch OR is invalid. However

Samsonovich et al (1992) simulated a microtubule as a 2-dimensional lattice plane with toroidal boundary conditions and found Fröhlich resonance maxima at discrete locations in super-lattice patterns on the simulated microtubule surface which precisely matched experimentally observed functional attachment sites for microtubule-associated proteins (MAPs). Further, Bandyopadhyay (2011) has experimental evidence for strong Fröhlich coherence in microtubules at multiple resonant frequencies.

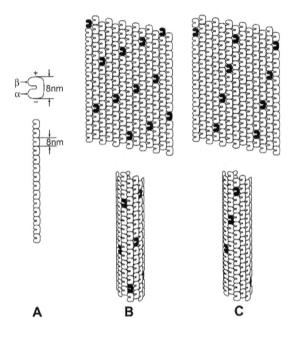

Figure 7. Simulating Fröhlich coherence in microtubules. A) Linear column of tubulins (protofilament) as simulated by Reimers et al (2010) which showed only weak Fröhlich condensation. B) and C) 2-dimensional tubulin sheets with toroidal boundary conditions (approximating 3-dimensional microtubule) simulated by Samsonovich et al (1992) shows long range Fröhlich resonance, with long-range symmetry, and nodes matching experimentally-observed MAP attachment patterns.

McKemmish et al (2010) challenged the Orch OR contention that tubulin switching is mediated by London forces, pointing out that mobile π electrons in a benzene ring (e.g. a phenyl ring without attachments) are completely delocalized, and hence cannot switch between states, nor exist in superposition of both states. Agreed. A single benzene cannot engage in switching. London forces occur between two or more electron cloud ring structures, or other non-polar groups. A single benzene ring cannot support London forces. It takes two (or more) to tango. Orch OR has always maintained two or more non-polar groups are necessary (Figure 8). McKemmish et al are clearly mistaken on this point.

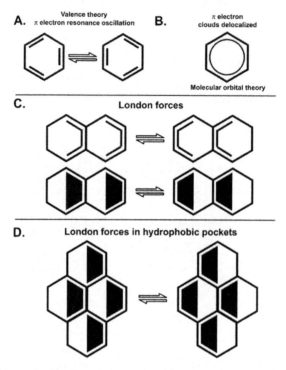

Figure 8. A) Phenyl ring/benzene of 6 carbons with three extra π electrons/ double bonds which oscillate between two configurations according to valence theory. B) Phenyl ring/benzene according to molecular orbital theory in which π electrons/double bonds are delocalized, thus preventing oscillation between alternate states. No oscillation/switching can occur. C) Two adjacent phenyl rings/benzenes in which π electrons/double bonds are coupled, i.e. van der Waals London (dipole dispersion) forces. Two versions are shown: In top version, lines represent double bond locations; in bottom version, dipoles are filled in to show negative charge locations. D) Complex of 4 rings with London forces.

McKemmish et al further assert that tubulin switching in Orch OR requires significant conformational structural change (as indicated in Figure 2), and that the only mechanism for such conformational switching is due to GTP hydrolysis, i.e. conversion of guanosine triphophate (GTP) to guanosine diphosphate (GDP) with release of phosphate group energy, and tubulin conformational flexing. McKemmish et al correctly point out that driving synchronized microtubule oscillations by hydrolysis of GTP to GDP and conformational changes would be prohibitive in terms of energy requirements and heat produced. This is agreed. However, we clarify that tubulin switching in Orch OR need not actually involve significant conformational change (e. g. as is illustrated in Figure 2), that electron cloud dipole states (London forces) are sufficient for bit-like switching, superposition and qubit function. We acknowledge tubulin conformational switching as discussed in early Orch OR publications and illustrations do indicate significant conformational changes. They are admittedly, though unintentionally, misleading.

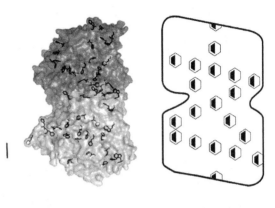

Figure 9. Left: Molecular simulation of tubulin with beta tubulin (dark gray) on top and alpha tubulin (light gray) on bottom. Non-polar amino acids phenylalanine and tryptophan with aromatic phenyl and indole rings are shown. (By Travis Craddock and Jack Tuszynski.) Right: Schematic tubulin with non-polar hydrophobic phenyl rings approximating actually phenyl and indole rings. Scale bar: 1 nanometer.

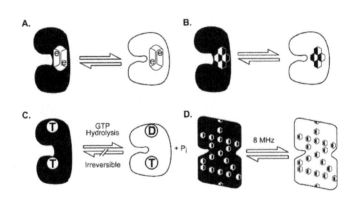

Figure 10. Four versions of the schematic Orch OR tubulin bit (superpositioned qubit states not shown). A) Early version showing conformational change coupled to/driven by single hydrophobic pocket with two aromatic rings. B) Updated version with single hydrophobic pocket composed of 4 aromatic rings. C) McKemmish et al (2009) mis-characterization of Orch OR tubulin bit as irreversible conformational change driven by GTP hydrolysis. D) Current version of Orch OR bit with no significant conformational change (change occurs at the level of atomic nuclei) and multiple hydrophobic pockets arranged in channels.

The only tubulin conformational factor in Orch OR is superposition separation involved in EG, the gravitational self-energy of the tubulin qubit. As previously described, we calculated EG for tubulin separated from itself at three possible levels: 1) the entire protein (e.g. partial separation, as suggested in Figure 2), 2) its atomic nuclei, and 3) its nucleons (protons and neutrons). The dominant effect is 2) separation at the level of atomic nuclei, e.g. 2.5 Fermi length for carbon nuclei (2.5 femtometers; 2.5 x 10-15 meters). This shift may be accounted for by London force dipoles with Mossbauer nuclear recoil and charge effects (Hameroff, 1998). Tubulin switching in Orch OR requires neither GTP hydrolysis nor significant conformational changes.

Schematic depiction of the tubulin bit, qubit and hydrophobic pockets in Orch OR has evolved over the years. An updated version is described in the next Section.

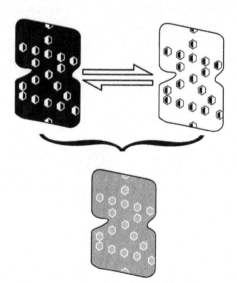

Figure 11. 2011 Orch OR tubulin qubit. Top: Alternate states of tubulin dimer (black and white) due to collective orientation of London force electron cloud dipoles in non-polar hydrophobic regions. There is no evident conformational change as suggested in previous versions; conformational change occurs at the level of atomic nuclei. Bottom: Depiction of tubulin (gray) superpositioned in both states.

9. Topological Quantum Computing in Orch OR

Quantum processes in Orch OR have consistently been ascribed to London forces in tubulin hydrophobic pockets, non-polar intra-protein regions, e.g. of π electron resonance rings of aromatic amino acids including tryptophan and phenylalanine. This assertion is based on (1) Fröhlich's suggestion that protein states are synchronized by electron cloud dipole oscillations in intra-protein non-polar regions, and (2) anesthetic gases selectively erasing consciousness by London forces in non-polar, hydrophobic regions in various neuronal proteins (e.g. tubulin, membrane proteins, etc.). London forces are weak, but numerous and able to act cooperatively to regulate protein states (Voet and Voet, 1995).

The structure of tubulin became known in 1998 (Nogales et al, 1998), allowing identification of non-polar amino acids and hydrophobic regions. Figure 9 shows locations of phenyl and indole π electron resonance rings of non-polar aromatic amino acids phenylalanine and tryptophan in tubulin. The ring locations are clustered along somewhat continuous pathways (within 2 nanometers) through tubulin. Thus, rather than hydrophobic pockets, tubulin may have within it quantum hydrophobic channels, or streams, linear arrays of electron resonance clouds suitable for cooperative, long-range quantum London forces. These quantum channels within each tubulin appear to align with those in adjacent tubulins in microtubule lattices, matching helical winding patterns (Figure 12). This in turn may support topological quantum computing in Orch OR.

Quantum bits, or qubits in quantum computers are generally envisioned as information bits in superposition of simultaneous alternative representations, e.g. both 1 and 0. Topological qubits are superpositions of alternative pathways, or channels which intersect repeatedly on a surface, forming 'braids'. Quasiparticles called anyons travel along such pathways, the intersections forming logic gates, with particular braids or pathways corresponding with particular information states, or bits. In superposition, anyons follow multiple braided pathways simultaneously, then reduce, or collapse to one particular pathway and functional output. Topological qubits are intrinsically resistant to decoherence.

An Orch OR qubit based on topological quantum computing specific to microtubule polymer geometry was suggested in Hameroff et al. (2002). Conductances along particular microtubule lattice geometry, e.g. Fibonacci helical pathways, were proposed to function as topological bits and qubits. Bandyopadhyay (2011) has preliminary evidence for ballistic conductance along different, discrete helical pathways in single microtubules

As an extension of Orch OR, we suggest topological qubits in microtubules based on quantum hydrophobic channels, e.g. continuous arrays of electron resonance rings within and among tubulins in microtubule lattices, e.g. following Fibonacci pathways. Cooperative London forces (electron cloud dipoles) in quantum hydrophobic channels may enable long-range coherence and topological quantum computing in microtubules necessary for brain function and consciousness.

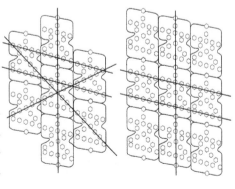

Figure 12. Left: Microtubule A-lattice configuration with lines connecting proposed hydrophobic channels of near-contiguous (<2 nanometer separation) electron resonance rings of phenylalanine and tryptophan. Right: Microtubule B-lattice with fewer such channels and lacking Fibonacci pathways. B-lattice microtubules have a vertical seam dislocation (not shown).

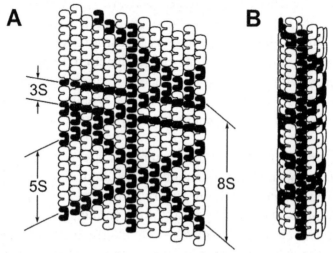

Figure 13. Extending microtubule A-lattice hydrophobic channels (Figure 12) results in helical winding patterns matching Fibonacci geometry. Bandyopadhyay (2011) has evidence for ballistic conductance and quantum inteference along such helical pathways which may be involved in topological quantum computing. Quantum electronic states of London forces in hydrophobic channels result in slight superposition separation of atomic nuclei, sufficient EG for Orch OR. This image may be taken to represent superposition of four possible topological qubits which, after time T=tau, will undergo OR, and reduce to specific pathway(s) which then implement function.

10. Conclusion: Consciousness in the Universe

Our criterion for proto-consciousness is OR . It would be unreasonable to refer to OR as the criterion for actual consciousness, because, according to the DP scheme, OR processes would be taking place all the time, and would be providing the effective randomness that is characteristic of quantum measurement. Quantum superpositions will continually be reaching the DP threshold for OR in non-biological settings as well as in biological ones, and usually take place in the purely random environment of a quantum system under measurement. Instead, our criterion for consciousness is Orch OR, conditions for which are fairly stringent: superposition must be isolated from the decoherence effects of the random environment for long enough to reach the DS threshold. Small superpositions are easier to isolate, but require longer reduction times τ. Large superpositions will reach threshold quickly, but are intrinsically more difficult to isolate. Nonetheless, we believe that there is evidence that such superpositions could occur within sufficiently large collections of microtubules in the brain for τ to be some fraction of a second.

Very large mass displacements can also occur in the universe in quantum-mechanical situations, for example in the cores of neutron stars. By OR, such superpositions would reduce extremely quickly, and classically unreasonable superpositions would be rapidly eliminated. Nevertheless, sentient creatures might have evolved in parts of the universe that would be highly alien to us. One possibility might be on neutron star surfaces, an idea that was developed ingeniously and in great detail by Robert Forward in two science-fiction stories (Dragon's Egg in 1980, Starquake in 1989). Such creatures (referred to as 'cheelas' in the books, with metabolic processes and OR-like events occurring at rates of around a million times that of a human being) could arguably have intense experiences, but whether or not this would be possible in detail is, at the moment, a very speculative matter. Nevertheless, the Orch OR proposal offers a possible route to rational argument, as to whether life of a totally alien kind such as this might be possible, or even probable, somewhere in the universe.

Such speculations also raise the issue of the 'anthropic principle', according to which it is sometimes argued that the particular dimensionless constants of Nature that we happen to find in our universe are 'fortuitously' favorable to human existence. (A dimensionless physical constant is a pure number, like the ratio of the electric to the gravitational force between the electron and the proton in a hydrogen atom, which in this case is a number of the general order of 10^{40}.) The key point is not so much to do with human existence, but the existence of sentient beings of any kind. Is there anything coincidental about the dimensionless physical constants being of such a nature that conscious life is possible at all? For example, if the mass of the neutron had been slightly less than that of the proton, rather than slightly larger, then neutrons rather than protons would have been stable, and this would be to the detriment of the whole subject of chemistry. These issues are frequently argued about (see Barrow and Tipler 1986), but the Orch OR proposal provides a little more substance to these arguments, since a proposal for the possibility of sentient life is, in principle, provided.

The recently proposed cosmological scheme of conformal cyclic cosmology (CCC) (Penrose 2010) also has some relevance to these issues. CCC posits that what we presently regard as the entire history of our universe, from its Big-Bang origin (but without inflation) to its indefinitely expanding future, is but one aeon in an unending succession of similar such aeons, where the infinite future of each matches to the big bang of the next via an infinite change of scale. A question arises whether the dimensionless constants of the aeon prior to ours, in the CCC scheme, are the same as those in our own aeon, and this relates to the question of whether sentient life could exist in that aeon as well as in our own. These questions are in principle answerable by observation, and again they would have a bearing on the extent or validity of the Orch OR proposal. If Orch OR turns

Cosmology of Consciousness

out to be correct, in it essentials, as a physical basis for consciousness, then it opens up the possibility that many questions may become answerable, such as whether life could have come about in an aeon prior to our own, that would have previously seemed to be far beyond the reaches of science.

Moreover, Orch OR places the phenomenon of consciousness at a very central place in the physical nature of our universe, whether or not this 'universe' includes aeons other than just our own. It is our belief that, quite apart from detailed aspects of the physical mechanisms that are involved in the production of consciousness in human brains, quantum mechanics is an incomplete theory. Some completion is needed, and the DP proposal for an OR scheme underlying quantum theory's R-process would be a definite possibility. If such a scheme as this is indeed respected by Nature, then there is a fundamental additional ingredient to our presently understood laws of Nature which plays an important role at the Planck-scale level of space-time structure. The Orch OR proposal takes advantage of this, suggesting that conscious experience itself plays such a role in the operation of the laws of the universe.

Acknowledgment We thank Dave Cantrell, University of Arizona Biomedical Communications for artwork.

References

Atema, J. (1973). Microtubule theory of sensory transduction. Journal of Theoretical Biology, 38, 181-90.

Bandyopadhyay A (2011) Direct experimental evidence for quantum states in microtubules and topological invariance. Abstracts: Toward a Science of Consciousness 2011, Sockholm, Sweden.

Barrow, J.D. and Tipler, F.J. (1986) The Anthropic Cosmological Principle (OUP, Oxford).

Bell, J.S. (1966) Speakable and Unspeakable in Quantum Mechanics (Cambridge Univ. Press, Cambridge; reprint 1987).

Benioff, P. (1982). Quantum mechanical Hamiltonian models of Turing Machines. Journal of Statistical Physics, 29, 515-46.

Bennett C.H., and Wiesner, S.J. (1992). Communication via 1- and 2-particle operators on Einstein-Podolsky-Rosen states. Physical Reviews Letters, 69, 2881-84.

Bensimon G, Chemat R (1991) Microtubule disruption and cognitive defects: effect of colchicine on teaming behavior in rats. Pharmacol. Biochem. Behavior 38:141-145.

Bohm, D. (1951) Quantum Theory (Prentice-Hall, Englewood-Cliffs.) Ch. 22, sect. 15-19. Reprinted as: The Paradox of Einstein, Rosen and Podolsky, in Quantum Theory and Measurement, eds., J.A. Wheeler and W.H. Zurek (Princeton University Press, Princeton, 1983).

Bernroider, G. and Roy, S. (2005) Quantum entanglement of K ions, multiple channel states and the role of noise in the brain. SPIE 5841-29:205-14.

Bouwmeester, D., Pan, J.W., Mattle, K., Eibl, M., Weinfurter, H. and Zeilinger, A. (1997) Experimental quantum teleportation. Nature 390 (6660): 575-579.

Brunden K.R., Yao Y., Potuzak J.S., Ferrer N.I., Ballatore C., James M.J., Hogan A.M., Trojanowski J.Q., Smith A.B. 3rd and Lee V.M. (2011) The characterization of microtubule-stabilizing drugs as possibletherapeutic agents for Alzheimer's disease and related taupathies. Pharmacological Research, 63(4), 341-51.

Chalmers, D. J., (1996). The conscious mind - In search of a fundamental theory. Oxford University Press, New York.

Crick, F., and Koch, C., (1990). Towards a neurobiological theory of consciousness. Seminars in the Neurosciences, 2, 263-75.

Dennett, D.C. (1991). Consciousness explained. Little Brown, Boston. MA.

Dennett, D.C. (1995) Darwin's dangerous idea: Evolution and the Meanings of Life, Simon and Schuster.

Dermietzel, R. (1998) Gap junction wiring: a 'new' principle in cell-to-cell communication in the nervous system? Brain Research Reviews. 26(2-3):176-83.

Deutsch, D. (1985) Quantum theory, the Church-Turing principle and the universal quantum computer, Proceedings of the Royal Society (London) A400, 97-117.

Diósi, L. (1987) A universal master equation for the gravitational violation of quantum mechanics, Physics Letters A 120 (8):377-381.

Diósi, L. (1989). Models for universal reduction of macroscopic quantum fluctuations Physical Review A, 40, 1165-74.

Draguhn A, Traub RD, Schmitz D, Jefferys (1998). Electrical coupling underlies high-frequency oscillations in the hippocampus in vitro. Nature, 394(6689), 189-92.

Eccles, J.C. (1992). Evolution of consciousness. Proceedings of the National Academy of Sciences, 89, 7320-24.

Engel GS, Calhoun TR, Read EL, Ahn T-K, Mancal T, Cheng Y-C, Blankenship RE, Fleming GR (2007) Evidence for wavelike energy transfer through quantum coherence in photosynthetic systems. Nature 446:782-786.

Everett, H. (1957). Relative state formulation of quantum mechanics. In Quantum Theory and Measurement, J.A. Wheeler and W.H. Zurek (eds.) Princeton University Press, 1983; originally in Reviews of Modern Physics, 29, 454-62.

Feynman, R.P. (1986). Quantum mechanical computers. Foundations of Physics, 16(6), 507-31.

Forward, R. (1980) Dragon's Egg. Ballentine Books.

Forward, R. (1989) Starquake. Ballentine Books.

Fröhlich, H. (1968). Long-range coherence and energy storage in biological systems. International Journal of Quantum Chemistry, 2, 641-9.

Fröhlich, H. (1970). Long range coherence and the actions of enzymes. Nature, 228, 1093.

Fröhlich, H. (1975). The extraordinary dielectric properties of biological materials and the action of enzymes. Proceedings of the National Academy of Sciences, 72, 4211-15.

Galarreta, M. and Hestrin, S. (1999). A network of fast-spiking cells in the neocortex connected by electrical synapses. Nature, 402, 72-75.

Gauger E., Rieper E., Morton J.J.L., Benjamin S.C., Vedral V. (2011) Sustained quantum coherence and entanglement in the avian compass http://arxiv.org/abs/0906.3725.

Ghirardi, G.C., Rimini, A., and Weber, T. (1986). Unified dynamics for microscopic and macroscopic systems. Physical Review D, 34, 470.

Ghirardi, G.C., Grassi, R., and Rimini, A. (1990). Continuous-spontaneous reduction model involving gravity. Physical Review A, 42, 1057-64.

Grush R., Churchland P.S. (1995), 'Gaps in Penrose's toilings', J. Consciousness Studies, 2 (1):10-29.

Hagan S, Hameroff S, and Tuszynski J, (2001). Quantum Computation in Brain Microtubules? Decoherence and Biological Feasibility, Physical Review E, 65, 061901.

Hameroff, S.R., and Watt R.C. (1982). Information processing in microtubules. Journal of Theoretical Biology, 98, 549-61.

Hameroff, S.R.(1987) Ultimate computing: Biomolecular consciousness and nanotechnology. Elsevier North-Holland, Amsterdam.

Hameroff, S.R., and Penrose, R., (1996a). Orchestrated reduction of quantum coherence in brain microtubules: A model for consciousness. In: Toward a Science of Consciousness ; The First Tucson Discussions and Debates. Hameroff, S.R., Kaszniak, and-Scott, A.C., eds., 507-540, MIT Press, Cambridge MA, 507-540. Also published in Mathematics and Computers in Simulation (1996) 40:453-480.

Hameroff, S.R., and Penrose, R. (1996b). Conscious events as orchestrated spacetime selections. Journal of Consciousness Studies, 3(1), 36-53.

Hameroff, S. (1998a). Quantum computation in brain microtubules?- The Penrose-Hameroff "Orch OR" model of consciousness. Philosophical Transactions of the Royal Society (London) Series A, 356, 1869-1896.

Hameroff, S. (1998b). 'Funda-mentality': is the conscious mind subtly linked to a basic level of the universe? Trends in Cognitive Science, 2, 119-127.

Hameroff, S. (1998c). Anesthesia, consciousness and hydrophobic pockets - A unitary quantum hypothesis of anesthetic action. Toxicology Letters, 100, 101, 31-39.

Hameroff, S. (1998d). HYPERLINK "http://www.hameroff.com/penrose-hameroff/cambrian.html"Did consciousness cause the Cambrian evolutionary explosion? In: Toward a Science of Consciousness II: The Second Tucson Discussions and Debates. Eds. Hameroff, S.R., Kaszniak, A.W., and Scott, A.C., MIT Press, Cambridge, MA.

Hameroff, S., Nip, A., Porter, M., and Tuszynski, J. (2002). Conduction pathways in microtubules, biological quantum computation and microtubules. Biosystems, 64(13), 149-68.

Hameroff S.R., & Watt R.C. (1982) Information processing in microtubules. Journal of Theoretical Biology 98:549-61.

Hameroff, S.R. (2006) The entwined mysteries of anesthesia and consciousness. Anesthesiology 105:400-412.

Hameroff, S.R, Craddock TJ, Tuszynski JA (2010) Memory 'bytes' - Molecular match for CaMKII phosphorylation encoding of microtubule lattices. Journal of Integrative Neuroscience 9(3):253-267.

He, R-H., Hashimoto, M., Karapetyan. H., Koralek, J.D., Hinton, J.P., Testaud, J.P., Nathan, V., Yoshida, Y., Yao, H., Tanaka, K., Meevasana, W., Moore, R.G., Lu, D.H.,Mo, S-K., Ishikado, M., Eisaki, H., Hussain, Z., Devereaux, T.P., Kivelson, S.A., Orenstein, Kapitulnik, J.A., Shen, Z-X. (2011) From a Single-Band Metal to a High Temperature Superconductor via Two Thermal Phase Transitions. Science, 2011;331 (6024): 1579-1583.

Hebb, D.O. (1949). Organization of Behavior: A Neuropsychological Theory, John Wiley and Sons, New York.

Huxley TH (1893; 1986) Method and Results: Essays.

Kant I (1781) Critique of Pure Reason (Translated and edited by Paul Guyer and Allen W. Wood, Cambridge University Press, 1998).

Kibble, T.W.B. (1981). Is a semi-classical theory of gravity viable? In Quantum Gravity 2: a Second Oxford Symposium; eds. C.J. Isham, R. Penrose, and D.W. Sciama (Oxford University Press, Oxford), 63-80.

Koch, C., (2004) The Quest for Consciousness: A Neurobiological Approach, Englewood, CO., Roberts and Co.

Koch C, Hepp K (2006) Quantm mechanics in the brain. Nature 440(7084):611.

Libet, B., Wright, E.W. Jr., Feinstein, B., & Pearl, D.K. (1979) Subjective referral of the timing for a conscious sensory experience. Brain 102:193-224.

Luo L, Lu J (2011) Temperature dependence of protein folding deduced from quantum transition. http://arxiv.org/abs/1102.3748

Lutz A, Greischar AL, Rawlings NB, Ricard M, Davidson RJ (2004) Long-term meditators self-induce high-amplitude gamma synchrony during mental practice The Proceedings of the National Academy of Sciences USA 101(46)16369-16373.

Macikic I., de Riedmatten H., Tittel W., Zbinden H. and Gisin N. (2002) Long-distance teleportation of qubits at telecommunication wavelengths Nature 421, 509-513.

Margulis, L. and Sagan, D. 1995. What is life? Simon and Schuster, N.Y.

Marshall, W, Simon, C., Penrose, R., and Bouwmeester, D (2003). Towards quantum superpositions of a mirror. Physical Review Letters 91, 13-16; 130401.

McKemmish LK, Reimers JR, McKenzie RH, Mark AE, Hush NS (2009) Penrose-Hameroff orchestrated objective-reduction proposal for human consciousness is not biologically feasible. Physical Review E. 80(2 Pt 1):021912.

Moroz, I.M., Penrose, R., and Tod, K.P. (1998) Spherically-symmetric solutions of the Schrödinger -Newton equations:. Classical and Quantum Gravity, 15, 2733-42.

Nogales E, Wolf SG, Downing KH. (1998) HYPERLINK "http://dx.doi.org/10.1038/34465"Structure of the $\alpha\beta$-tubulin dimer by electron crystallography. Nature. 391, 199-203.

Ouyang, M., & Awschalom, D.D. (2003) Coherent spin transfer between molecularly bridged quantum dots. Science 301:1074-78.

Pearle, P. (1989). Combining stochastic dynamical state-vector reduction with spontaneous localization. Physical Review A, 39, 2277-89.

Pearle, P. and Squires, E.J. (1994). Bound-state excitation, nucleon decay experiments and models of wave-function collapse. Physical Review Letters,

73(1), 1-5.

Penrose, R. (1989). The Emperor's New Mind: Concerning Computers, Minds, and the Laws of Physics, Oxford University Press, Oxford.

Penrose, R. (1993). Gravity and quantum mechanics. In General Relativity and Gravitation 13. Part 1: Plenary Lectures 1992. Proceedings of the Thirteenth International Conference on General Relativity and Gravitation held at Cordoba, Argentina, 28 June - 4 July 1992. Eds. R.J.Gleiser, C.N.Kozameh, and O.M.Moreschi (Inst. of Phys. Publ. Bristol and Philadelphia), 179-89.

Penrose, R. (1994). Shadows of the Mind; An Approach to the Missing Science of Consciousness. Oxford University Press, Oxford.

Penrose, R. (1996). On gravity's role in quantum state reduction. General Relativity and Gravitation, 28, 581-600.

Penrose, R. (2000). Wavefunction collapse as a real gravitational effect. In Mathematical Physics 2000, Eds. A.Fokas, T.W.B.Kibble, A.Grigouriou, and B.Zegarlinski. Imperial College Press, London, 266-282.

Penrose, R. (2002). John Bell, State Reduction, and Quanglement. In Quantum Unspeakables: From Bell to Quantum Information, Eds. Reinhold A. Bertlmann and Anton ZeilingeR, Springer-Verlag, Berlin, 319-331.

Penrose, R. (2004). The Road to Reality: A Complete Guide to the Laws of the Universe. Jonathan Cape, London.

Penrose, R. (2009). Black holes, quantum theory and cosmology (Fourth International Workshop DICE 2008), Journal of Physics, Conference Series 174, 012001.

Penrose, R. (2010). Cycles of Time: An Extraordinary New View of the Universe. Bodley Head, London.

Penrose R. and Hameroff S.R. (1995) What gaps? Reply to Grush and Churchland. Journal of Consciousness Studies.2:98-112.

Percival, I.C. (1994) Primary state diffusion. Proceedings of the Royal Society (London) A, 447, 189-209.

Pokorn-, J., Hasek, J., Jel-nek, F., Saroch, J. & Palan, B. (2001) Electromagnetic activity of yeast cells in the M phase. Electro Magnetobiol 20, 371-396.

Pokorn-, J. (2004) Excitation of vibration in microtubules in living cells. Bioelectrochem. 63: 321-326.

Polkinghorne, J. (2002) Quantum Theory, A Very Short Introduction. Oxford University Press, Oxford.

Rae, A.I.M. (1994) Quantum Mechanics. Institute of Physics Publishing; 4th edition 2002.

Rasmussen, S., Karampurwala, H., Vaidyanath, R., Jensen, K.S., and Hameroff, S. (1990) Computational connectionism within neurons: A model of cytoskeletal automata subserving neural networks. Physica D 42:428-49.

Reimers JR, McKemmish LK, McKenzie RH, Mark AE, Hush NS (2009) Weak, strong, and coherent regimes of Frohlich condensation and their applications to terahertz medicine and quantum consciousness Proceedings of the National Academy of Sciences USA 106(11):4219-24

Reiter GF, Kolesnikov AI, Paddison SJ, Platzman PM, Moravsky AP, Adams MA, Mayers J (2011) Evidence of a new quantum state of nano-confined water http://arxiv.org/abs/1101.4994

Rieper E, Anders J, Vedral V (2011) Quantum entanglement between the electron clouds of nucleic acids in DNA. http://arxiv.org/abs/1006.4053.

Samsonovich A, Scott A, Hameroff S (1992) Acousto-conformational transitions in cytoskeletal microtubules: Implications for intracellular information processing. Nanobiology 1:457-468.

Sherrington, C.S. (1957) Man on His Nature, Second Edition, Cambridge University Press.

Smolin, L. (2002). Three Roads to Quantum Gravity. Basic Books. New York.

Tegmark, M. (2000) The importance of quantum decoherence in brain processes. Physica Rev E 61:4194-4206.

Tittel, W, Brendel, J., Gisin, B., Herzog, T., Zbinden, H., and Gisin, N. (1998) Experimental demonstration of quantum correlations over more than 10 km, Physical Review A, 57:3229-32.

Tononi G (2004) An information integration theory of consciousness BMC Neuroscience 5:42.

Turin L (1996) A spectroscopic mechanism for primary olfactory reception Chem Senses 21(6) 773-91.

Tuszynski JA, Brown JA, Hawrylak P, Marcer P (1998) Dielectric polarization, electrical conduction, information processing and quantum computation in microtubules. Are they plausible? Phil Trans Royal Society A 356:1897-1926.

Tuszynski, J.A., Hameroff, S., Sataric, M.V., Trpisova, B., & Nip, M.L.A. (1995) Ferroelectric behavior in microtubule dipole lattices; implications for information processing, signaling and assembly/disassembly. J. Theoretical Biology 174:371-80.

Voet, D., Voet, J.G. 1995. Biochemistry, 2nd edition. Wiley, New York.

von Rospatt, A., (1995) The Buddhist Doctrine of Momentariness: A survey of the origins and early phase of this doctrine up to Vasubandhu (Stuttgart: Franz Steiner Verlag).

Wegner, D.M. (2002) The illusion of conscious will Cambridge MA, MIT Press.

Whitehead, A.N., (1929) Process and Reality. New York, Macmillan.

Whitehead, A.N. (1933) Adventure of Ideas, London, Macmillan.

Wigner E.P. (1961). Remarks on the mind-body question, in The Scientist Speculates, ed. I.J. Good (Heinemann, London). In Quantum Theory and Measurement, eds., J.A. Wheeler and W.H. Zurek, Princeton Univsity Press, Princeton, MA. (Reprinted in E. Wigner (1967), Symmetries and Reflections, Indiana University Press, Bloomington).

Wolfram, S. (2002) A New Kind of Science. Wolfram Media incorporated.

What Consciousness Does:
A Quantum Cosmology of Mind

Chris J. S. Clarke, Ph.D.

School of Mathematics, University of Southampton, University Road,
Southampton SO17 1BJ, UK

Abstract

This article presents a particular theoretical development related to the conceptualisation of the role of consciousness by Hameroff and Penrose. The first three sections review, respectively: the different senses of "consciousness" and the sense to be used in this article; philosophical conceptions of how consciousness in this sense can be said to do anything; and the historical development of understanding of the role of consciousness in quantum theory. This background is then drawn upon in the last two sections, which present a cosmological perspective in which consciousness and quantum theory are complementary processes governed by different logics.

1. Consciousness: What Are We Talking About?

"Consciousness" is notoriously difficult to define because it is so fundamental; it is the precondition for our being able to do or know anything. Surveying the voluminous controversies over the meaning of the word suggests, however, that "consciousness" tends to be used in two fairly distinct ways. Broadly considered, just as the word "spirit" has two quite different referential meanings, viz., alcoholic beverage vs religion/parapsychology, so "consciousness" has two meanings: One meaning refers essentially to subjective experience: our moment- by-moment qualitative awareness of what is happening both internally (thoughts, feelings) and externally. It is the "what it is like" of Nagel's seminal paper (Nagel, 1974). The other meaning is that used by Dennett (1991). Taking to heart Wittgenstein's dictum that "whereof one cannot speak, thereof one must be silent", he restricts the topic of consciousness to those aspects of experience that we can report on, verbally, to other people. From this he is led to restrict the concept to that part of our internal experience that is contained in our inner dialogue, the almost constant talking to ourselves whereby we make sense of the world to ourselves in verbal terms. Thus consciousness in Dennett's sense of the word is the process of our forming "drafts" of parts of our internal dialogue. I shall call these senses of "consciousness" as used by Nagel and Dennett qualis-consciousness and quid-consciousness, respectively, from the Latin words for "how" and "what". Qualis is related to quale, quality, and with the idea that qualis-consciousness is comprised of qualities (qualia) associated with perception and thinking.

This difference between these two senses is crucial when we consider what we know about the consciousness of other people or other beings. Whereas we can, and by its definition can only, explore the quid-consciousness of another person by talking to them, we can only know the non-verbal part of another's qualis-consciousness through empathy; that is, through our evolved capacity for mirroring the sensations of others in response to a range of bodily cues and contextual information (Berger, 1987). This means that, as cogently argued by Nagel (1974), in the case of an organism like a bat with which it is difficult to have much empathy, we cannot know explicitly that they have qualis-consciousness, even though we might postulate that this is the case because they are mammals like ourselves. On the other hand when it comes to the verbally based quid-consciousness we know that bats, lacking language, cannot have it. The distinction between the two concepts is thus vital when discussing non-human consciousness. Without deeper analysis, we cannot rule out the occurrence of a "hidden" qualis-consciousness from any organism, or even from physical systems that we may not consider organisms at all - a vital point that will be revisited in section 5.

2. What Does Consciousness Do?

A major strand in the philosophy of consciousness concerns the notion of epiphenomenalism - the idea that consciousness is an add-on that appears upon ("epi") information processing without having any functional role. This is not relevant to quid-consciousness, which is actually a part of information processing rather than something added to it. In the case of qualis-consciousness, on the other hand, "epiphenomenalism" seems meaningful while at the same time seeming odd, because the whole notion of causation, in the physical sense, seems problematic in connection with qualis-consciousness. This consciousness does not do things like digesting food or moving limbs in a purely mechanical sense, but it comprises the whole of our experienced world (McGilchrist, 2009) and thereby establishes the context and preconditions as a result of which doing-events like moving limbs take place. The distinction between the two senses of "consciousness" lies not in whether they either do things or not; it lies in the distinct categories of "doing" and "being" that are involved. Quid-consciousness does things in a causal sense as part of a whole control structure of information processing. Qualis-consciousness constitutes a meaningful world within which doing is possible.

Here it becomes a matter of one's philosophical position, whether or not qualis-consciousness is anything other than a sort of emotional fog generated by processing in the brain. If one adopts a scientific-realist position on which the world is entirely reducible to mechanical processes, then qualis-consciousness is indeed such a fog. The alternative to this is to recognise that there is a whole area of discourse concerning existence, value, meaning and so on which is related to the mechanical properties of the world, but which is not equivalent to the mechanical aspect of the world.

Since it is qualis-consciousness that raises the most significant problems in consciousness studies, I shall from now on restrict attention to this sense of the word and, with this understood, I shall usually drop the "qualis" and just call it "consciousness".

3. The Changing View of the Role of Consciousness in Quantum Mechanics.

The earlier history of this topic falls into four phases:

(a) The "quantum theory" of Planck and Einstein, based on a conventionally mechanical concept of "quanta".

(b) The "quantum mechanics" of Bohr and Heisenberg from about 1925. This was based on complementarity and the uncertainty principle, which increasingly involved the idea of the collapse of the quantum state (also known as the wave function). It culminated in von Neumann's picture (von Neumann, 1932) of two quite distinct processes: a smooth deterministic evolution of the state under a dynamics, and a discontinuous transition from one state to another related to observation. He did not, however, suppose that consciousness was peculiarly concerned in this, arguing that it was sufficient to consider the human being as an assemblage of rather sensitive physical detectors.

(c) The views of Wigner and London and Bauer (London & Bauer, 1939, 1983) that consciousness was essential for collapse. According to this, the quantum state of the human brain was, through the process of experimental observation, coupled to the quantum state of a microscopic system; then the consciousness of the human being collapsed the joint state of human, apparatus and microsystem. This role for consciousness was strictly limited. Consciousness was not responsible for determining what particular quantity was being measured, because this was determined by the apparatus (a point that will be revisited in section 5). It could not bias the probabilities for different outcomes, because this would undermine the very laws of physics. All that consciousness could do was, somehow, to demand that some definite outcome did emerge, rather than a mixture or superposition of possibilities.

(d) A focus on the quantum-classical distinction. This began with Daneri, Loinger and Prosperi (1962) suggesting that "collapse" was the transition from a quantum state to a classical state, and that this was located not in the brain of the observer, but in the experimental apparatus. They showed that it was the large size of the apparatus, with a large number of possible quantum states all linked to the state of the microsystem being observed, which averaged out the peculiarly quantum mechanical nature of the microsystem, resulting in an essentially classical, non-quantum state for the apparatus. Subsequently Zeh (1970) included in this averaging-out of quantum states the highly effective role of interaction with the wider universe through the phenomenon of "decoherence". By this time the idea that consciousness had a role in quantum theory came to be regarded as superfluous.

In consequence of this history, it has become clear that we are here dealing with two distinct (though interrelated) physical representations. One is the superposition of states, a peculiarly quantum effect resulting in, for instance, the interference patterns produced in the experiment where particles are fired towards two parallel slits. The other is the statistical mixture of states used to represent mathematically a situation such as the result of a rolling a dice, where there is a range of possible outcomes with different probabilities for each. Considered purely mathematically, decoherence turns a superposition into a mixture. This does not, however, explain why we are actually aware, at the end of the process, of one particular outcome as opposed to a fuzzy blur of possibilities. We may recall that this, and only this, was what consciousness was supposed to achieve on London and Bauer's earlier way of looking at things. Despite much clarification between 1939 and 1970, the possible role for consciousness has remained little changed, and its operational details have until recently remained obscure.

4. The Perspective of Cosmology on the Role of Consciousness

More recent arguments from the surprising direction of cosmology now clarify things a great deal. In particular, quantum cosmology starkly underlines the need for something like consciousness. To take a particular example: the WMAP satellite observations of the universe at an age of some 380,000 years confirm a picture in which the universe has evolved as if it started in a perfectly smooth homogenous state (though strictly speaking there can be no "initial state" since the very earliest stages merge into the as yet unknown timeless conditions of quantum gravity). By the epoch observed by WMAP we see minute fluctuations superimposed on this uniform background, of the same character as the quantum fluctuations that can be detected when a uniform beam of radiation is observed in the laboratory. On conventional theory, these cosmological fluctuations grew under the influence of gravity to produce stars, galaxies and ourselves. Note, however, that in quantum theory it is the act of observation that precipitates quantum fluctuations: without observation (in whatever generalised form we may conceive it) a homogeneous initial state evolving under homogeneous laws must remain homogeneous. So the early fluctuations that eventually give rise to the existence of planets, people and WMAP are caused by observations such as those made by people and WMAP! The problem of quantum observation lies at the heart of modern cosmology.

This cosmological perspective makes it clear that the bare mathematical formalism of quantum theory in insufficient on its own. Without some additional ingredient, the universe would remain homogeneous and sterile. Two ideas from quantum cosmology are needed in order to make sense of this. They will also provide the key to the role of consciousness.

The first was introduced by James Hartle (1991), building on the "histories" interpretation of Griffiths (1984). Instead of considering probabilities for different outcomes to a single quantum observation, Hartle examined the probabilities of sets of outcomes for any collection of observations scattered throughout the universe in space and time. The mathematics was almost the same as it would have been if one had assumed a collapse of the wave function simultaneously across the universe with each observation; but strictly speaking the latter concept cannot be used in cosmology because it is not consistent with the fact that in relativity theory "simultaneous" is an observer-dependent concept. By considering this "super- observation" extended over the whole of space-time there is no need to consider either collapse or issues of causality between future and past events.

Hartle gave no indication as to what was actually meant by an "observation" or "observer". This issue was made explicit through the second key idea, first raised by Matthew Donald. He considered quid-consciousness - i.e. information processing - but this cannot help because it is in no way essentially different from any other purely mechanical process. Then, however, the idea was explored by Don Page who focussed on "sensation", which is close to the qualis-consciousness of this paper. The aim of a cosmological theory, he argued, was to explain the universe as we see it, and this is equivalent to requiring that the quantum state of the universe is compatible with an instance of conscious sensation like ours. This in turn is equivalent to the quantum state assigning a non-zero probability to such an instance. This then gives a new way of thinking about the role of consciousness: consciousness does not alter the quantum state of the universe, but it imposes a filter on the state, selecting a component (if there is one) compatible with our capacity for sensation.

The combined work of Hartle and Page gives a picture of a universe arising from the interplay of a background homogeneous quantum cosmology with possible networks in space and time of instances of consciousness. Self-contradictory networks of awareness are ruled out because quantum mechanics assigns to them a zero probability (Everett, 1957; Clarke, 1974) . But in addition the networks of awareness are shaped by their own internal logic, manifested by qualis-consciousness and different from the Aristotelian logic of quid-consciousness (Clarke, 2007). This logic brings in elements such as agency and meaning. Consciousness, on this view, "does something", but by selection rather than modification, and in a way which is compatible with and dependent on the known laws of physics.

5. A Theoretical Understanding of Consciousness and Quantum Theory

One final building block still seems required: a non-arbitrary criterion is needed for what physical systems have the capacity for (qualis-)consciousness. Many recent authors (de Quincey, 2002; Skrbina, 2005) have, however, come to the conclusion that no such criterion exists. In other words, everything might be conscious, a position known as "panpsychism". A problem remains, however: if "everything" is conscious, what is a "thing"? The answer of Heidegger (1967) concerned only a pejorative cultural aspect of the word; the answer of Döring and Isham (2011) invokes an ad hoc external mechanism; instead we need to explore naturally occurring physical criteria for what is a thing. A consciousness-carrying "thing" must have some internal unity rather than being an arbitrary aggregate of objects, which suggests that it has an internal coherence. The simplest definition of this is that its parts are in quantum entanglement (Clarke, 2007) . In addition, it must not be merely an arbitrary subset of a larger "thing", so that it must be maximal with respect to this coherence. In other words, it must be on the boundary between the quantum and the classical, a boundary set by the onset of decoherence. The structures considered by Hameroff and Penrose (1996) are of this sort.

It now becomes clearer what consciousness does. At this quantum-classical boundary the question of what "observation" (or, more formally, what algebra of propositions) is to be expressed is not yet determined by decoherence, and so is open to determination by consciousness (Clarke, 2007). Following Hartle, this happens not in isolation, but within the whole network of "things" throughout the universe. Physical causation operates through quantum state of the universe, while consciousness independently filters this into awareness through its own sort of logic (in the sense of the structure of an algebra of propositions). The large scope this gives for future experimental and theoretical research has been outlined in (Clarke, 2007,8). Several candidates for the logic of consciousness are available, allowing us to understand how consciousness brings creativity alongside rational deduction. This model raises for the first time the possibility of a rigorous theoretical framework for parapsychology (Clarke, 2008) without which that subject remains only a semi-science. It turns out that consciousness can itself, through the "Zeno effect", enlarge the length scale for the onset of decoherence, which then offers hope for understanding how small-scale elements can be "orchestrated", in Hameroff's sense (Hammeroff & Penrose, 1996), into the ego-consciousness known to us. In addition, there will be other candidates for what a "thing" is, opening up alternative theories that can be tested against the theory just outlined.

References

Berger, D. M. (1987). Clinical empathy. Northvale: Jason Aronson, Inc.

Clarke, C. J. S. (1974) Quantum Theory and Cosmology. Philosophy of Science, 41, 317-332.

Clarke, C. J. S. (2007). The role of quantum physics in the theory of subjective consciousness. Mind and Matter 5(1), 45-81.

Clarke C. J. S. (2008). A new quantum theoretical framework for parapsychology. European Journal of Parapsychology, 23(1), 3-30.

Daneri, A., Loinger, A., Prosperi, G. M. (1962). Quantum Theory of Measurement and Ergodicity Conditions., Nuclear Physics 33, 297-319.

de Quincey, C. (2002). Radical Nature: Rediscovering the Soul of Matter. Montpelier VT : Invisible Cities Press.

Dennett, D. C., (1991). Consciousness Explained, Allen Lane.

Donald, M. (1990). Quantum Theory and the Brain, Proceedings of the Royal Society (London) Series A, 427, 43-93.

Döring, A., Isham, C. (2011). "What is a Thing?": Topos Theory in the Foundations of Physics. In B. Coecke (Ed.), New Structures for Physics, Lecture Notes in Physics, Vol. 813 (pp 753-941). Berlin: Springer.

Everett, H., (1957). Relative State Formulation of Quantum Mechanics, Reviews of Modern Physics 29, pp 454-462.

Griffiths, R..B. (1984). Consistent histories and the interpretation of quantum mechanics. J. Stat. Phys. 36, 219-272.

Hameroff, S., Penrose, R. (1996). Conscious events as orchestrated space-time selections. Journal of Consciousness Studies, 3(1), 36-53.

Hartle, J. (1991). The quantum mechanics of cosmology. In Coleman, S., Hartle, P., Piran, T., Weinberg, S., (Eds) Quantum cosmology and baby universes. Singapore: World Scientific.

Heidegger M. (1967). What Is a Thing? (Trans. W. B. Barton Jr., V. Deutsch). Chicago: Henry Regnery Company.

London, F., Bauer, E. (1939). La théorie de l'observation en mécanique quantique. Hermann, Paris.

London, F., Bauer, E. (1983). The theory of observation in quantum mechanics (translation of the above). In J. A. Wheeler, W. H. Zurek (Eds), Quantum Theory and Measurement (pp. 217- 259). Princeton: Princeton University Press.

McGilchrist, I. (2009). The Master and his Emissary: the divided brain and the making of the Western world. New Haven and London: Yale University Press.

Nagel, T. (1974). What Is it Like to Be a Bat? Philosophical Review 83(4), 435-450.

Skrbina, D. (2005). Panpsychism in the West. Cambridge MA: Bradford Books.

von Neumann, J. (1932). Mathematical Foundations of Quantum Mechanics, (Beyer, R. T., trans.), Princeton Univ. Press.

Zeh, H. D., (1970). On the Interpretation of Measurement in Quantum Theory, Foundation of Physics, 1, pp. 69-76.

Quantum Physics and the Multiplicity of Mind: Split-Brains, Fragmented Minds, Dissociation, Quantum Consciousness

Rhawn Joseph, Ph.D.,

Emeritus, Brain Research Laboratory, Northern California.

Abstract

Quantum physics and Einstein's theory of relativity make assumptions about the nature of the mind which is assumed to be a singularity. In the Copenhagen model of physics, the process of observing is believed to effect reality by the act of perception and knowing which creates abstractions and a collapse function thereby inducing discontinuity into the continuum of the quantum state. This gives rise to the uncertainty principle. Yet neither the mind or the brain is a singularity, but a multiplicity which include two dominant streams of consciousness and awareness associated with the left and right hemisphere, as demonstrated by patients whose brains have been split, and which are superimposed on yet other mental realms maintained by the brainstem, thalamus, limbic system, and the occipital, temporal, parietal, and frontal lobes. Like the quantum state, each of these minds may also become discontinuous from each other and each mental realm may perceive their own reality. Illustrative examples are detailed, including denial of blindness, blind sight, fragmentation of the body image, phantom limbs, the splitting of the mind following split-brain surgery, and dissociative states where the mind leaves the body and achieves a state of quantum consciousness and singularity such that the universe and mind become one.

1. Introduction

In 1905 Albert Einstein published his theories of relativity, which promoted the thesis that reality and its properties, such as time and motion had no objective "true values", but were "relative" to the observer's point of view (Einstein, 1905a,b,c). However, what if the observer is not a singularity and has more than one point of view and more than one stream of observing consciousness? And what if these streams of consciousness were also relative?

Cosmology of Consciousness

Quantum physics, as exemplified by the Copenhagen school (Bohr, 1934, 1958, 1963; Heisenberg, 1930, 1955, 1958), also makes assumptions about the nature of reality as related to an observer, the "knower" who is conceptualized as a singularity. Because the physical world is relative to being known by a "knower" (the observing consciousness), then the "knower" can influence the nature of the reality which is being observed. In consequence, what is known vs what is not known becomes relatively imprecise (Heisenberg, 1958).

For example, as expressed by the Heisenberg uncertainty principle (Heisenberg, 1955, 1958), the more precisely one physical property is known the more unknowable become other properties, whose measurements become correspondingly imprecise. The more precisely one property is known, the less precisely the other can be known and this is true at the molecular and atomic levels of reality. Therefore it is impossible to precisely determine, simultaneously, for example, both the position and velocity of an electron.

However, we must ask: if knowing A, makes B unknowable, and if knowing B makes A unknowable, wouldn't this imply that both A and B, are in fact unknowable? If both A and B are manifestations of the processing of "knowing," and if observing and measuring can change the properties of A or B, then perhaps both A and B are in fact properties of knowing, properties of the observing consciousness, and not properties of A or B.

In quantum physics, nature and reality are represented by the quantum state. The electromagnetic field of the quantum state is the fundamental entity, the continuum that constitutes the basic oneness and unity of all things.

The physical nature of this state can be "known" by assigning it mathematical properties (Bohr, 1958, 1963). Therefore, abstractions, i.e., numbers, become representational of a hypothetical physical state. Because these are abstractions, the physical state is also an abstraction and does not possess the material consistency, continuity, and hard, tangible, physical substance as is assumed by Classical (Newtonian) physics. Instead, reality, the physical world, is created by the process of observing, measuring, and knowing (Heisenberg, 1955).

Consider an elementary particle, once this positional value is assigned, knowledge of momentum, trajectory, speed, and so on, is lost and becomes "uncertain." The particle's momentum is left uncertain by an amount inversely proportional to the accuracy of the position measurement which is determined by values assigned by the observing consciousness. Therefore, the nature of reality, and the uncertainty principle is directly affected by the observer and the process of observing and knowing (Heisenberg, 1955, 1958).

Quantum Physics, Neuroscience of Mind

The act of knowing creates a knot in the quantum state; described as a "collapse of the wave function;" a knot of energy that is a kind of blemish in the continuum of the quantum field. This quantum knot bunches up at the point of observation, at the assigned value of measurement.

The process of knowing, makes reality, and the quantum state, discontinuous. "The discontinuous change in the probability function takes place with the act of registration...in the mind of the observer" (Heisenberg, 1958).

Reality, therefore, is a manifestation of alterations in the patterns of activity within the electromagnetic field which are perceived as discontinuous. The perception of a structural unit of information is not just perceived, but is inserted into the quantum state which causes the reduction of the wave-packet and the collapse of the wave function.

Knowing and not knowing, are the result of interactions between the mind and concentrations of energy that emerge and disappear back into the electromagnetic quantum field.

However, if reality is created by the observing consciousness, and can be made discontinuous, does this leave open the possibility of a reality behind the reality? Might there be multiple realities? And if consciousness and the observer and the quantum state is not a singularity, could each of these multiple realities also be manifestations of a multiplicity of minds?

Heinsenberg (1958) recognized this possibility of hidden realities, and therefore proposed that the reality that exists beyond or outside the quantum state could be better understood when considered in terms of "potential" reality and "actual" realities. Therefore, although the quantum state does not have the ontological character of an "actual" thing, it has a "potential" reality; an objective tendency to become actual at some point in the future, or to have become actual at some point in the past.

Therefore, it could be said that the subatomic particles which make up reality, or the quantum state, do not really exist, except as probabilities. These "subatomic" particles have probable existences and display tendencies to assume certain patterns of activity that we perceive as shape and form. Yet, they may also begin to display a different pattern of activity such that being can become nonbeing and thus something else altogether.

The conception of a deterministic reality is therefore subjugated to mathematical probabilities and potentiality which is relative to the mind of a knower which effects that reality as it unfolds, evolves, and is observed (Bohr 1958, 1963; Heisenberg 1955, 1958). That is, the mental act of perceiving a non-localized unit of structural information, injects that mental event into the quantum state of the universe, causing "the collapse of the wave function" and creating a bunching up, a tangle and discontinuous knot in the continuity of the quantum state.

Einstein ridiculed these ideas (Pais, 1979): "Do you really think the moon isn't there if you aren't looking at it?"

Heisenberg (1958), cautioned, however, that the observer is not the creator of reality: "The introduction of the observer must not be misunderstood to imply that some kind of subjective features are to be brought into the description of nature. The observer has, rather, only the function of registering decisions, i.e., processes in space and time, and it does not matter whether the observer is an apparatus or a human being; but the registration, i.e., the transition from the "possible" to the "actual," is absolutely necessary here and cannot be omitted from the interpretation of quantum theory."

Shape and form are a function of our perception of dynamic interactions within the continuum which is the quantum state. What we perceive as mass (shape, form, length, weight) are dynamic patterns of energy which we selectively attend to and then perceive as stable and static, creating discontinuity within the continuity of the quantum state. Therefore, what we are perceiving and knowing, are only fragments of the continuum.

However, we can only perceive what our senses can detect, and what we detect as form and shape is really a mass of frenzied subatomic electromagnetic activity that is amenable to detection by our senses and which may be known by a knowing mind. It is the perception of certain aspects of these oscillating patterns of continuous evolving activity, which give rise to the impressions of shape and form, and thus discontinuity, as experienced within the mind. This energy that makes up the object of our perceptions, is therefore but an aspect of the electromagnetic continuum which has assumed a specific pattern during the process of being sensed and processed by those regions of the brain and mind best equipped to process this information. Perceived reality, therefore, becomes a manifestation of mind.

However, if the mind is not a singularity, and if we possessed additional senses or an increased sensory channel capacity, we would perceive yet other patterns and other realities which would be known by those features of the mind best attuned to them. If the mind is not a singularity but a multiplicity, this means that both A and B, may be known simultaneously.

2. Duality vs Multiplicity

In the Copenhagen model, the observer is external to the quantum state the observer is observing, and they are not part of the collapse function but a witness of it (Bohr, 1958, 1963; Heisenberg 1958). However, if the Copenhagen model is correct, and as the cosmos contains observers, then the standard collapse formulation can not be used to describe the entire universe as the universe contains observers (von Neumann, 1932, 1937).

Further, reality becomes, at a minimum, a duality (observer and observed) with the potential to become a multiplicity.

As described by DeWitt and Graham (1973; Dewitt, 1971), "This reality, which is described jointly by the dynamical variables and the state vector, is not the reality we customarily think of, but is a reality composed of many worlds. By virtue of the temporal development of the dynamical variables the state vector decomposes naturally into orthogonal vectors, reflecting a continual splitting of the universe into a multitude of mutually unobservable but equally real worlds, in each of which every good measurement has yielded a definite result and in most of which the familiar statistical quantum laws hold."

The minimal duality is that aspect of reality which is observed, measured, and known, and that which is unknown. However, this minimal duality is an illusion as indicated not only by the potential to become multiplicity, but by the nature of mind which is not a singularity (Joseph, 1982, 1986a; 1988a,b).

Even if we disregard the concept of "mind" and substitute the word "brain", the fact remains that the brain is not a singularity. The human brain is functionally specialized with specific functions and different mental states localized to specific areas, each of which is capable of maintaining independent and semi-independent aspects of conscious-awareness (Joseph 1986a,b, 1988a,b, 1992, 1999a). Different aspects of the same experience and identical aspects of that experience may be perceived and processed by different brain areas in different ways (Gallagher and Joseph, 1982; Joseph 1982; Joseph and Gallagher 1985; Joseph et al., 1984).

Therefore, although it has been said that orthodox quantum mechanics is completely concordant with the defining characteristics of Cartesian dualism, this is an illusion. Cartesian duality assumes singularity of mind, when in fact, the overarching organization of the mind- and the brain- is both dualistic and multiplistic.

If quantum physics is "mind-like" (actual/operational at the quantum level, but mentalistic on the ontological level) then quantum physics, or rather, the quantum state (reality, the universe) is not a duality, but a multiplicity. Indeed, the entire concept of duality is imposed on reality by the dominant dualistic nature of the brain and mind which subordinates not just reality, but the multiplicity of minds maintained within the human brain (Joseph, 1982).

Like the Copenhagen school, Von Neumann's formulation of quantum mechanics (1932, 1937), fails to recognize or understand the multiple nature of mind and reality. Von Neumann postulated that the physical aspects of nature are represented by a density matrix. The matrix, therefore, could be conceptualized as a subset of potential realities, and that by averaging the values of these evolving matrices, the state of the universe and thus of reality, can be ascertained as a unified whole. However, in contrast to the Copenhagen interpretation, Von Neumann shifted the observer (his brain) into the quantum universe and thus made it subject to the rules of quantum physics.

Ostensibly and explicitly, Von Neumann's conceptions are based on a conception of mind as a singularity acting on the quantum state which contains the brain. Von Neumann's mental singularity, therefore, imposes itself on reality, such that each "event" that occurs within reality, is associated with one specific experience of the singularity-mind. Thus, Von Neumann assumes the brain and mind has only "one experience" which corresponds with "one event;" and this grossly erroneous misconception of the nature of the brain and mind, unfortunately, is erroneously accepted as fact by most cosmologists and physicists. Further, he argues that in the process of knowing, the quantum state of this singularity brain/mind also collapses, or rather, is reduced in a mathematically quantifiable manner, just as the quantum universe is collapsed and reduced by being known (Von Neumann, 1932, 1937).

However, the brain and mind are not a singularity, but a multiplicity (Joseph, 1982, 1988a,b, 1999a). Nevertheless, Von Neumann's conceptions can be applied to the multiplicity of mind/brain when each mental realm is considered individually as an interactional subset of the multiplicity.

3. The Multiplicity of Mind and Perception

According to Von Neumann (1932), the "experiential increments in a person's knowledge" and "reductions of the quantum mechanical state of that person's brain" corresponds to the elimination of all those perceptual functions that are not necessary or irrelevant to the knowing of the event and the increase in the knowledge associated with the experience.

If considered from the perspective of an isolated aspect of the mind and the dominating stream of consciousness, Von Neuman's conceptions are essentially correct. However, neither the brain nor the mind function in isolation but in interaction with other neural tissues and mental/perceptual/sensory realms (Joseph 1982, 1992, 1999a). Perceptual functions are not "eliminated" and removed from the brain. Instead, they are prevented from interfering with the attentional processes of one aspect of the multiplicity of mind which dominates during the knowing event (Joseph, 1986b, 1999a).

Consider, by way of example, you are sitting in your office reading this text. The pressure of the chair, the physical sensations of your shoes and clothes, the musculature of your body as it holds one then another position, the temperature of the room, various odors and fragrances, a multitude of sounds, visual sensations from outside your area of concentration and focus, and so on, are all being transmitted to the brainstem, midbrain, and olfactory limbic system. These signals are then relayed to various subnuclei within the thalamus.

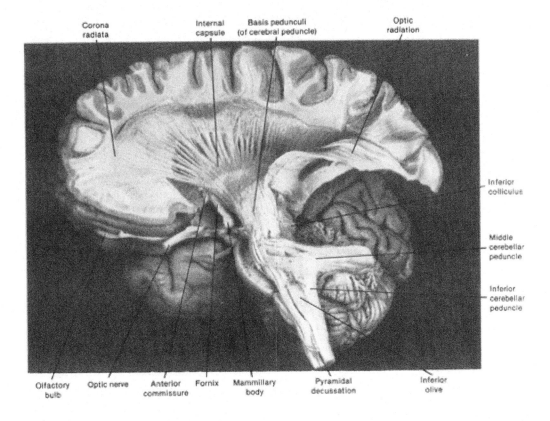

The neural tissues of the brainstem, midbrain, limbic system and thalamus are associated with the "old brain." However, those aspects of consciousness we most closely associated with humans are associated with the "new brain"

the neocortex (Joseph, 1982, 1992). Therefore, although you may be "aware" of these sensations while they are maintained within the old brain, you are not "conscious" of them, unless a decision is made to become conscious or they increase sufficiently in intensity that they are transferred to the neocortex via the thalamus and frontal lobes, and forced into the focus of consciousness (Joseph, 1982, 1986b, 1992, 1999a).

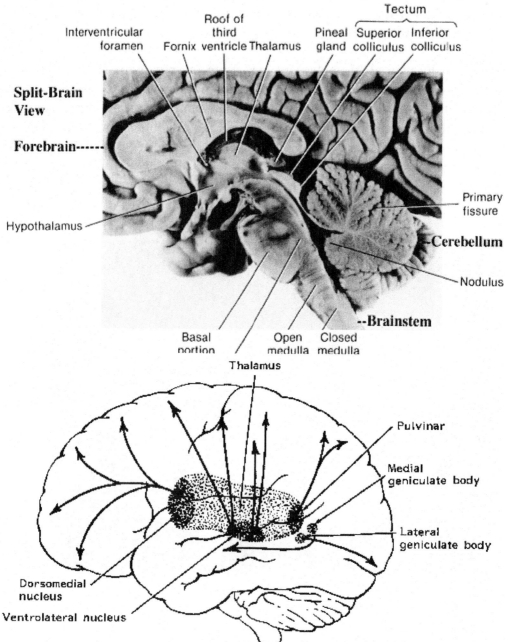

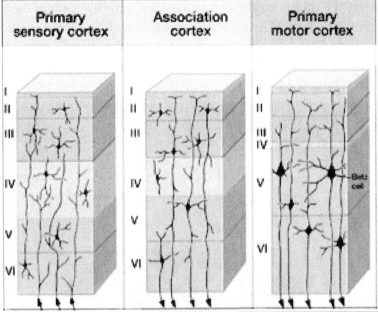

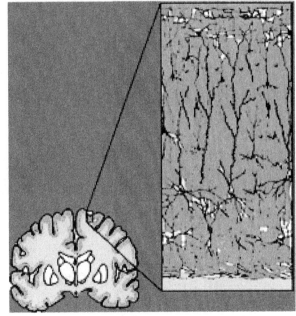

The old brain is covered by a gray mantle of new cortex, neocortex. The sensations alluded to are transferred from the old brain to the thalamus which relays these signals to the neocortex. Human consciousness and the "higher" level of the multiplicity of mind, are associated with the "new brain."

For example, visual input is transmitted from the eyes to the midbrain and thalamus and is transferred to the primary visual receiving area maintained in the neocortex of the occipital lobe (Casagrande & Joseph 1978, 1980; Joseph and Casagrande, 1978). Auditory input is transmitted from the inner ears to the brainstem, midbrain, and thalamus, and is transferred to the primary auditory receiving area within the neocortex of the temporal lobe. Tactual-physical stimuli are also transmitted from the thalamus to the primary somatosensory areas maintained in the neocortex of the parietal lobe. From the primary areas these signals are transferred to the adjoining "association" areas, and simple percepts become more complex by association (Joseph, 1996).

Monitoring all this perceptual and sensory activity within the thalamus and neocortex is the frontal lobes of the brain, also known as the senior executive of the brain and personality (Joseph 1986b, 1999a; Joseph et al., 1981). It is the frontal lobes which maintain the focus of attention and which can selectively inhibit any additional processing of signals received in the primary areas.

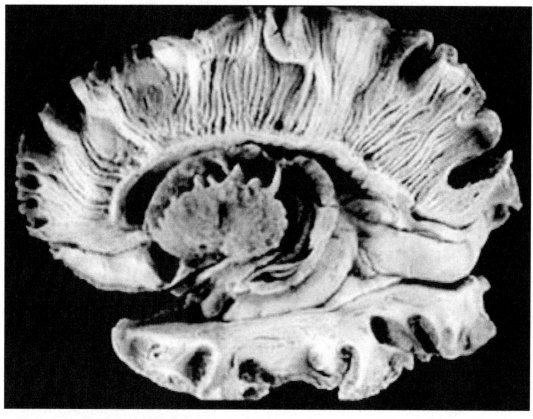

Figures: The Corona Radiata: Neural pathways linking to the neocortex

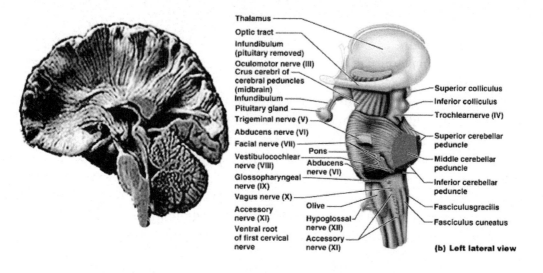

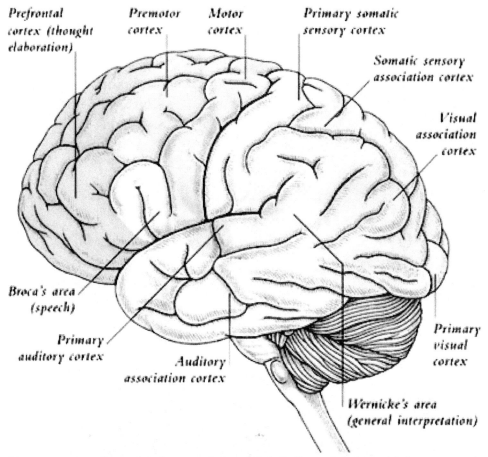

There are two frontal lobes, a right and left frontal lobe which communicate via a bridge of nerve fibers. Each frontal lobe, and subdivisions within each are concerned with different types of mental activity (Joseph, 1999a).

The left frontal lobe, among its many functions, makes possible the ability to speak. It is associated with the verbally expressive, speaking aspects of consciousness. However, there are different aspects of consciousness associated not only with the frontal lobe, but with each lobe of the brain and its subdivisions (Joseph, 1986b; 1996, 1999a).

4. Knowing Yet Not Knowing: Disconnected Consciousness

Consider the well known phenomenon of "word finding difficulty" also known as "tip of the tongue." You know the word you want (the "thingamajig") but at the same time, you can't gain access to it. That is, one aspect of consciousness knows the missing word, but another aspect of consciousness associated with talking and speech can't gain access to the word. The mind is disconnected from itself. One aspect of mind knows, the other aspect of mind does not.

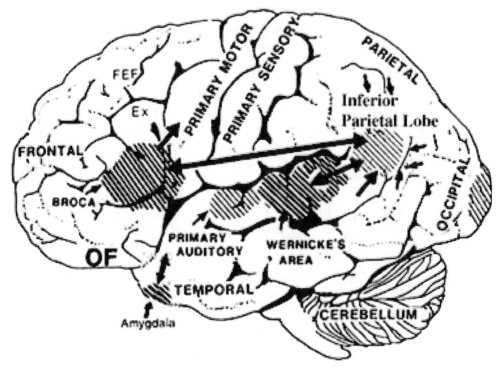

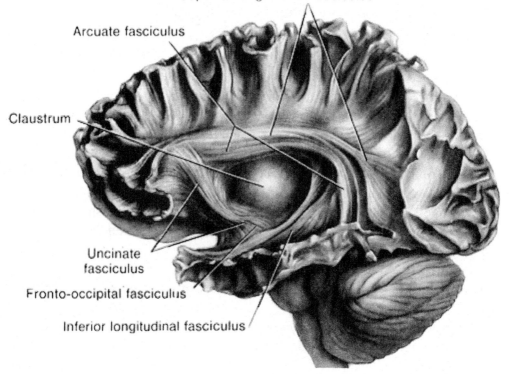

This same phenomenon, but much more severe and disabling, can occur if the nerve fiber pathway linking the language areas of the left hemisphere are damaged. For example, Broca's area in the frontal lobe organizes words received from the posterior language areas, and expresses humans speech. Wernicke's area in the temporal lobe comprehends speech. The inferior parietal lobe in association with the frontal lobes and Broca's area, associates and assimilates associations so that, for example, we can say the word "dog" and come up with the names of dozens of different breeds and then visualize and describe them (Joseph, 1982; Joseph and Gallagher 1985; Joseph et al., 1984). Therefore, if Broca's area is disconnected from the posterior language areas, one aspect of consciousness may know what it wants to say, but the speaking aspect of consciousness will be unable to gain access to it and will have nothing to say; called "conduction aphasia."

Or consider damage which disconnects the parietal lobe from Broca's area. If you place an object, e.g., a comb, out-of-sight, in the person's right hand, and ask them to name the object, the speaking aspect of consciousness may know something is in the hand, but will be unable to name it. However, although they can't name it, and can't guess if shown pictures, if the patient is asked to point to the correct object, they will correctly pick out the comb (Joseph, 1996). Therefore, part of the brain and mind may act purposefully (e.g. picking out the comb), whereas another aspect of the brain and mind is denied access to the information that the disconnected part of the mind is acting on.

Thus, the part of the brain and mind which is perceiving and knowing, is not the same as the part of the brain and mind which is speaking. This phenomenon occurs even in undamaged brains, when the multiplicity of minds which make up one of the dominant streams of consciousness, become disconnected and/or are unable to communicate.

5. The Visual Mind: Denial of Blindness

All visual sensations first travel from the eyes to the thalamus and midbrain. At this level, these visual impressions are outside of consciousness, though we may be aware of them. These visual sensations are then transferred to the primary visual receiving areas and to the adjacent association areas in the neocortex of the occipital lobe. Once these visual impressions reach the neocortex, consciousness of the visual word is achieved. Visual consciousness is made possible by the occipital lobe.

Destruction of the occipital lobe and its neocortical visual areas results in cortical blindness (Joseph, 1996). The consciousness mind is blinded and can not see or sense anything except vague sensations of lightness and darkness. However, because visual consciousness is normally maintained within the occipital lobe, with destruction of this tissue, the other mental systems will not know that they can't see. The remaining mental system do not know they are blind.

Wernicke's area in the left temporal lobe in association with the inferior parietal lobe comprehends and can generate complex language. Normally, visual input is transferred from the occipital to the inferior parietal lobe (IPL) which is adjacent to Wernicke's area and the visual areas of the occipital lobe. Once these signals arrive in the IPL a person can name what they see; the visual input is matched with auditory-verbal signals and the conscious mind can label and talk about what is viewed (Joseph, 1982, 1986b; Joseph et al., 1984). Talking and verbally describing what is seen is made possible when this stream of information is transferred to Broca's area in the left frontal lobe (Joseph, 1982, 1999a). It is Broca's area which speaks and talks.

Therefore, with complete destruction of the occipital lobe, visual consciousness is abolished whereas the other mental system remain intact but are unable to receive information about the visual world. In consequence, the verbal aspects of consciousness and the verbal-language mind does not know it can't see because the brain area responsible for informing these mental system about seeing, no longer exists. . In fact the language-dependent conscious mind will deny that it is blind; and this is called: Denial of blindness.

Normally, if it gets dark, or you close your eyes, the visual mind becomes conscious of this change in light perception and will alert the other mental realms. These other mental realms do not process visual signals and therefore they must be informed about what the visual mind is seeing. If the occipital lobe is destroyed, visual consciousness is destroyed, and the rest of the brain cannot be told that visual consciousness can't see. Therefore, the rest of the brain does not know it is blind, and when asked, will deny blindness and will make up reasons for why they bump into furniture or can't recognize objects held before their eyes (Joseph, 1986b, 1988a).

For example, when unable to name objects, they might confabulate an explanation: "I see better at home." Or, "I tripped because someone moved the furniture."

Even if you tell them they are blind, they will deny blindness; that is, the verbal aspects of consciousness will claim it can see, when it can't. The Language-dependent aspects of consciousness does not know that it is blind because information concerning blindness is not being received from the mental realms which support visual consciousness.

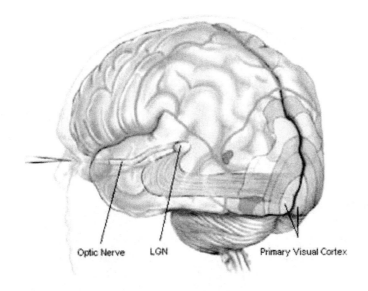

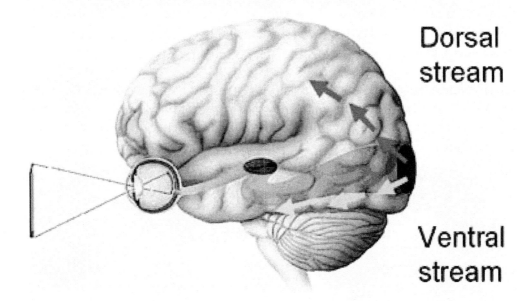

The same phenomenon occurs with small strokes destroying just part of the occipital lobe. Although a patient may lose a quarter or even half of their visual field, they may be unaware of it. This is because that aspect of visual consciousness no longer exists and can't inform the other mental realms of its condition.

6. "Blind Sight"

The brains of reptiles, amphibians, and fish do not have neocortex. Visual input is processed in the midbrain and thalamus and other old-brain areas as

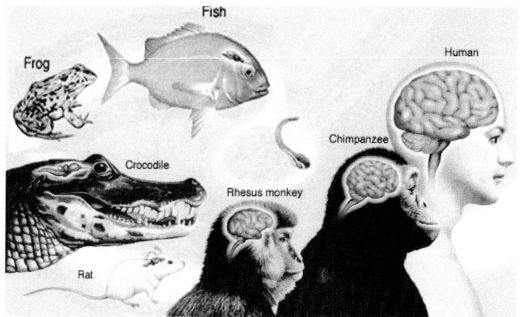

these creatures do not possess neocortex or lobes of the brain. In humans, this information is also received in the brainstem and thalamus and is then transferred to the newly evolved neocortex.

As is evident in non-mammalian species, these creatures can see, and they are aware of their environment. They possess an older-cortical (brainstem-thalamus) visual awareness which in humans is dominated by neocortical visual consciousness

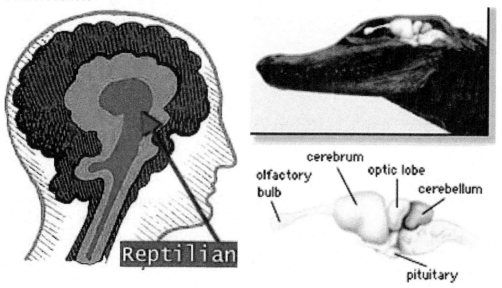

Figures: (Left) Human Brain. Reptile Brain (Right)

Therefore, even with complete destruction of the visual neocortex, and after the patient has had time to recover, some patients will demonstrate a non-conscious awareness of their visual environment. Although they are cortically blind and can't name objects and stumble over furniture and bump into walls, they may correctly indicate if an object is moving in front of their face, and they may turn their head or even reach out their arms to touch it--just as a frog can see a fly buzzing by and lap it up with its tongue. Although the patient can't name or see what has moved in front of his face, he may report that he has a "feeling" that something has moved.

Frogs do not have neocortex and they do not have language, and can't describe what they see. However, humans and frogs have old cortex that process visual impressions and which can control and coordinate body movements. Therefore, although the neocortical realms of human consciousness are blind, the mental realms of the old brain can continue to see and can act on what it sees; and this is called: Blind sight (Joseph, 1996).

7. Body Consciousness: Denial of the Body, and Phantom Limbs

All tactile and physical-sensory impressions are relayed from the body to the brainstem and the thalamus, and are then transferred to the primary receiving and then the association area for somatosensory information located in the neocortex of the parietal lobe (Joseph, 1986b, 1996). The entire image of the body is represented in the parietal lobes (the right and left half of the body in the left and right parietal lobe respectively), albeit in correspondence with the sensory importance of each body part. Therefore, more neocortical space is devoted to the hands and fingers than to the elbow.

It is because the body image and body consciousness is maintained in the parietal area of the brain, that victims of traumatic amputation and who lose an arm or a leg, continue to feel as if their arm or their leg is still attached to the body. This is called: phantom limbs. They can see the leg is missing, but they feel as if it is still there; body-consciousness remains intact even though part of the body is missing (Joseph, 1986b, 1996). They may also continue to periodically experience the pain of the physical trauma which led to the amputation, and this is called "phantom limb pain."

Thus, via the mental system of the parietal lobe, consciousness of what is not there, may appear to consciousness as if it is still there. This is not a hallucination. The image of the body is preserved in the brain and so to is consciousness of the body; and this is yet another example of experienced reality being a manifestation of the brain and mind. In this regard, reality is literally mapped into the brain and is represented within the brain, such that even when aspects of this "reality" are destroyed and no longer exists external to the brain, it nevertheless continues to be perceived and experienced by the brain and the associated realms of body-consciousness.

Conversely, if the parietal lobe is destroyed, particularly the right parietal lobe (which maintains an image of the left half of the body), half of the body image may be erased from consciousness (Joseph, 1986b, 1988a). The remaining realms of mind will lose all consciousness of the left half of the body, which, in their minds, never existed.

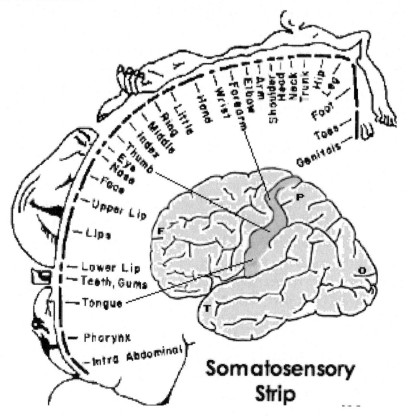

Doctor: "Give me your right hand!" (Patient offers right hand). "Now give me your left!" (The patient presents the right hand again. The right hand is held.) "Give me your left!" (The patient looks puzzled and does not move.) "Is there anything wrong with your left hand?"

Patient: "No, doctor."

Doctor: "Why don't you move it, then?"

Patient: "I don't know."

Doctor: "Is this your hand?" (The left hand is held before her eyes.)

Patient: "Not mine, doctor."

Doctor: "Whose hand is it, then?"

Patient: "I suppose it's yours, doctor."

Doctor: "No, it's not; I've already got two hands. Look at it carefully." (The left hand is again held before her eyes.)

Patient: "It is not mine, doctor."

Doctor: "Yes it is, look at that ring; whose is it?" (Patient's finger with marriage ring is held before her eyes)

Patient: "That's my ring; you've got my ring, doctor. You're wearing my ring!"

Doctor: "Look at it—it is your hand."

Patient: "Oh, no doctor."

Doctor: "Where is your left hand then?"

Patient: "Somewhere here, I think." (Making groping movements near her left shoulder).

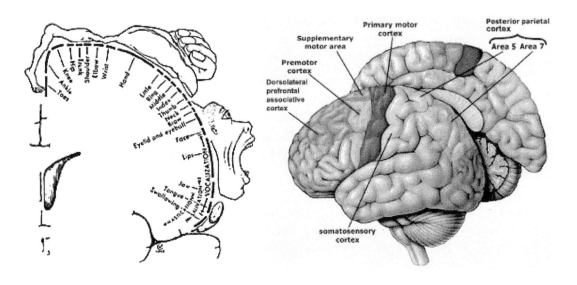

Cosmology of Consciousness

Because the body image has been destroyed, consciousness of that half of the body is also destroyed. The remaining mental systems and the language-dependent conscious mind will completely ignore and fail to recognize their left arm or leg because the mental system responsible for consciousness of the body image no longer exists. If the left arm or leg is shown to them, they will claim it belongs to someone else, such as the nurse or the doctor. They may dress or groom only the right half of their body, eat only off the right half of their plates, and even ignore painful stimuli applied to the left half of their bodies (Joseph, 1986b, 1988a).

However, if you show them their arm and leg (whose ownership they deny), they will admit these extremities exists, but will insist the leg or arm does not belong to them, even though the arm or the leg is wearing the same clothes covering the rest of their body. Instead, the language dependent aspects of consciousness will confabulate and make up explanations and thus create their own reality. One patient said the arm belonged to a little girl, whose arm had slipped into the patient's sleeve. Another declared (speaking of his left arm and leg), "That's an old man. He stays in bed all the time."

One such patient engaged in peculiar erotic behavior with his left arm and leg which he believed belonged to a woman. Some patients may develop a dislike for their left arms, try to throw them away, become agitated when they are referred to, entertain persecutory delusions regarding them, and even complain of strange people sleeping in their beds due to their experience of bumping into their left limbs during the night (Joseph, 1986b, 1988a). One patient complained that the person sharing her bed, tried to push her out of the bed and then insisted that if it happened again she would sue the hospital. Another complained about "a hospital that makes people sleep together." A female patient expressed not only anger but concern least her husband should find out; she was convinced it was a man in her bed.

The right and left parietal lobes maintain a map and image of the left and right half of the body, respectively. Therefore, when the right parietal lobe is destroyed, the language-dependent mental systems of the left half of the brain, having access only to the body image for the right half of the body, is unable to become conscious of the left half of their body, except as body parts that they then deduce must belong to someone else.

However, when the language dominant mental system of the left hemisphere denies ownership of the left extremity these mental system are in fact telling the truth. That is, the left arm and leg belongs to the right not the left hemisphere; the mental system that is capable of becoming conscious of the left half of their body no longer exist.

When the language axis (Joseph, 1982, 2000), i.e. the inferior parietal lobe, Broca's and Wernicke's areas, are functionally isolated from a particular source of information, the language dependent aspect of mind begins to make up a response based on the information available. To be informed about the left leg or left arm, it must be able to communicate with the cortical area (i.e. the parietal lobe) which is responsible for perceiving and analyzing information regarding the extremities. When no message is received and when the language axis is not informed that no messages are being transmitted, the language zones instead rely on some other source even when that source provides erroneous input (Joseph, 1982, 1986b; Joseph et al., 1984); substitute material is assimilated and expressed and corrections cannot be made (due to loss of input from the relevant knowledge source). The patient begins to confabulate. This is because the patient who speaks to you is not the 'patient' who is perceiving- they are in fact, separate; multiple minds exist in the same head.

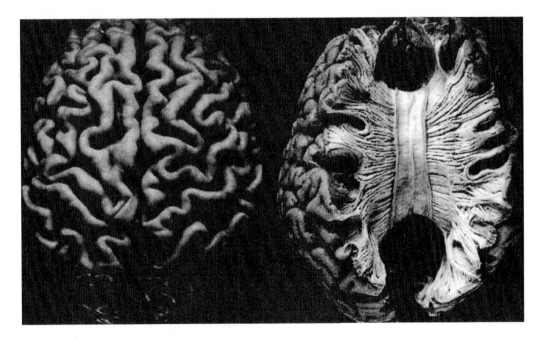

8. Split-Brains and Split-Minds.

The multiplicity of mind is not limited to visual consciousness, body consciousness, or the language-dependent consciousness. Rather the multiplicity of mind include social consciousness, emotional consciousness, and numerous other mental realms linked with specific areas of the brain (Joseph, 1982, 1986a,b, 1988a,b, 1992, 1999a), such as the limbic system (emotion), frontal lobes (rational thought), the inferior temporal lobes (memory) and the two halves of the brain where multiple streams of mental activity become subordinated and dominated by two distinct realms of mind; consciousness and awareness (Joseph,

1982, 1986a,b, 1988a,b).

The brain is not a singularity. This is most apparent when viewing the right and left half of the brain which are divided by the interhemispheric fissure and almost completely split into two cerebral hemispheres. These two brain halves are connected by a rope of nerve fibers, the corpus callosum, which enables them to share and exchange some information, but not all information as these two mental realms maintain a conscious awareness of different realities.

For example, it is well established that the right cerebral hemisphere is dominant over the left in regard to the perception, expression and mediation of almost all aspects of social and emotional functioning and related aspects of social/ emotional language and memory. Further, the right hemisphere is dominant for most aspects of visual-spatial perceptual functioning, the comprehension of body language, the recognition of faces including friend's loved ones, and one's own face in the mirror (Joseph, 1988a, 1996).

Recognition of one's own body and the maintenance of the personal body image is also the dominant realm of the right half of the brain (Joseph, 1986b, 1988a). The body image, for many, is tied to personal identity; and the same is true of the recognition of faces including one's own face.

The right is also dominant for perceiving and analyzing visual-spatial relationships, including the movement of the body in space (Joseph, 1982, 1988a). Therefore, one can throw or catch a ball with accuracy, dance across a stage, or leap across a babbling brook without breaking a leg.

The perception of environmental sounds (water, wind, a meowing cat) and the social, emotional, musical, and melodic aspects of language, including the ability to sing, curse, or pray, are also the domain of the right hemisphere mental system (Joseph, 1982, 1988a). Hence, it is the right hemisphere which imparts the sounds of sarcasm, pride, humor, love, and so on, into the stream of speech, and which conversely can determine if others are speaking with sincerity, irony, or evil intentions.

By contrast, expressive and receptive speech, linguistic knowledge and thought, mathematical and analytical reasoning, reading, writing, and arithmetic, as well as the temporal-sequential and rhythmical aspects of consciousness, are associated with the functional integrity of the left half of the brain in the majority of the population (Joseph, 1982, 1996). The language-dependent mind is linked to the left hemisphere.

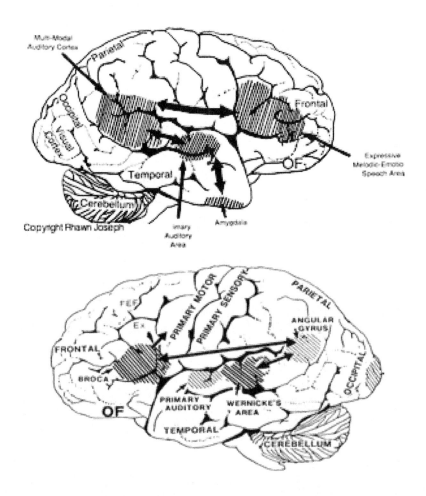

Certainly, there is considerable overlap in functional representation. Moreover, these two mental system interact and assist the other, just as the left and right hands cooperate and assist the other in performing various tasks. For example, if you were standing at the bar in a nightclub, and someone were to tap you on the shoulder and say, "Do you want to step outside," it is the mental system of the left hemisphere which understands that a question about "outside" has been asked, but it is the mental system of the right which determines the underlying meaning, and if you are being threatened with a punch in the nose, or if a private conversation is being sought.

However, not all information can be transferred from the right to the left, and vice versa (Gallagher and Joseph, 1982; Joseph, 1982, 1988a; Joseph and Gallagher, 1985; Joseph et al., 1984). Because each mental system is unique, each "speaks a different language" and they cannot always communicate. Not all mental events can be accurately translated, understood, or even recognized by the other half of the brain. These two major mental systems, which could be likened to "consciousness" vs "awareness" exist in parallel, simultaneously, and both can act independently of the other, have different goals and desires, and come to completely different conclusions. Each mental system has its own reality.

The existence of these two independent mental realms is best exemplified and demonstrated following "split-brain" surgery; i.e. the cutting of the corpus callosum fiber pathway which normally allows the two hemisphere's to communicate.

As described by Nobel Lauriate Roger Sperry (1966, p. 299), "Everything we have seen indicates that the surgery has left these people with two separate minds, that is, two separate spheres of consciousness. What is experienced in the right hemisphere seems to lie entirely outside the realm of awareness of the left hemisphere. This mental division has been demonstrated in regard to perception, cognition, volition, learning and memory."

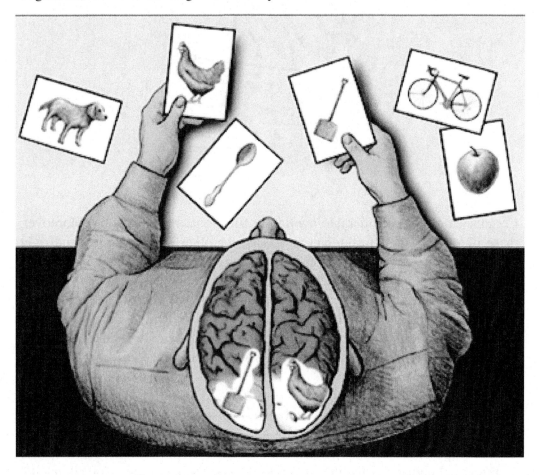

The right half of the brain controls and perceives the left half of the body and visual space, whereas the right half of the body and visual space is the domain of the left hemisphere. Therefore, following split-brain surgery, if a comb, spoon, or some other hidden object is placed in the left hand (out of sight), the left hemisphere, and the language-dependent conscious mind, will not even know the left hand is holding something and will be unable to name it, describe it, or if given multiple choices point to the correct item with the right hand (Joseph 1988a,b; Sperry, 1966). However, the right hemisphere can raise the left hand and not only point to the correct object, but can pantomime its use.

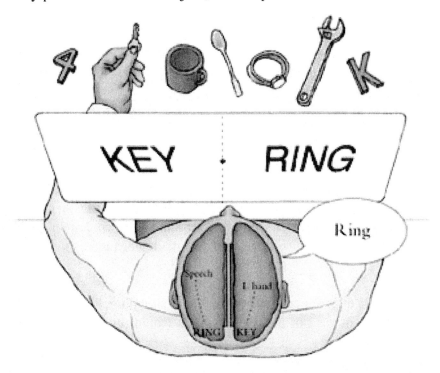

If the split-brain patient is asked to stare at the center of a white screen and words like "Key Ring" are quickly presented, such that the word "Key" falls in the left visual field (and thus, is transmitted to the right cerebrum) and the word "Rings" falls in the right field (and goes to the left hemisphere), the language dependent conscious mind will not see the word "Key." If asked, the language-dependent conscious mind will say "Ring" and will deny seeing the word "Key." However, if asked to point with the left hand, the mental system of the right hemisphere will correctly point to the word "Key."

Therefore, given events "A" and "B" one half of the brain may know A, but know nothing about B which is known only by the other half of the brain. In consequence, what is known vs what is not known becomes relatively imprecise depending on what aspects of reality are perceived and "known" by which mental system (Joseph 1986a, 1988a). There is no such thing as singularity of mind. Since the brain and mind is a multiplicity, "A" and "B" can be known simultaneously, even when one mind is knows nothing about the existence of A or B.

In that the brain of the normal as well as the "split-brain" patient maintains the neuroanatomy to support a multiplicity of mind, and the presence of two dominant psychic realms, it is therefore not surprising that "normal" humans often have difficulty "making up their minds," suffer internal conflicts over love/ hate relationships, and are plagued with indecision even when staring into an open refrigerator and trying to decide what to eat. "Making up one's mind" can be an ordeal involving a multiplicity of minds.

However, this conflict becomes even more apparent following split-brain surgery and the cutting of the corpus callosum fiber pathway which links these two parallel streams of conscious-awareness.

Akelaitis (1945, p. 597) describes two patients with complete corpus callosotomies who experienced extreme difficulties making the two halves of their bodies cooperate. "In tasks requiring bimanual activity the left hand would frequently perform oppositely to what she desired to do with the right hand. For example, she would be putting on clothes with her right and pulling them off with her left, opening a door or drawer with her right hand and simultaneously pushing it shut with the left. These uncontrollable acts made her increasingly irritated and depressed."

Another patient experienced difficulty while shopping, the right hand would place something in the cart and the left hand would put it right back again and grab a different item.

A recently divorced male patient complained that on several occasions while walking about town he found himself forced to go some distance in another direction by his left leg. Later (although his left hemisphere was not conscious of it at the time) it was discovered that this diverted course, if continued, would have led him to his former wife's new home.

Geschwind (1981) reports a callosal patient who complained that his left hand on several occasions suddenly struck his wife--much to the embarrassment of his left (speaking) hemisphere. In another case, a patient's left hand attempted to choke the patient himself and had to be wrestled away.

Bogen (1979, p. 333) indicates that almost all of his "complete commissurotomy patients manifested some degree of intermanual conflict." One patient, Rocky, experienced situations in which his hands were uncooperative; the right would button up a shirt and the left would follow right behind and undo the buttons. For years, he complained of difficulty getting his left leg to go in the direction he (or rather his left hemisphere) desired. Another patient often referred to the left half of her body as "my little sister" when she was complaining of its peculiar and independent actions.

Another split-brain patient reported that once when she had overslept her left hand began slapping her face until she (i.e. her left hemisphere) woke up. This same patient, in fact, complained of several instances where her left hand had acted violently toward herself and other people (Joseph, 1988a).

Split brain patient, 2-C, complained of instances in which his left hand would perform socially inappropriate actions, such as striking his mother across the face (Joseph, 1988b). Apparently his left and right hemisphere also liked different TV programs. He complained of numerous instances where he (his left hemisphere) was enjoying a program, when, to his astonishment, the left half of his body pulled him to the TV, and changed the channel.

The right and left hemisphere also liked different foods and had different attitudes about exercise. Once, after 2-C had retrieved something from the refrigerator with his right hand, his left took the food, put it back on the shelf and retrieved a completely different item "Even though that's not what I wanted to eat!" On at least one occasion, his left leg refused to continue "going for a walk" and would only allow him to return home.

In the laboratory, 2-C's left hemisphere often became quite angry with his left hand, and he struck it and expressed hate for it. Several times, his left and right hands were observed to engage in actual physical struggles, beating upon each other. For example, on one task both hands were stimulated simultaneously (while out of view) with either the same or two different textured materials (e.g., sandpaper to the right, velvet to the left), and he was required to point (with the left and right hands simultaneously) to an array of fabrics that were hanging in view on the left and right of the testing apparatus. However, at no time was he informed that two different fabrics were being applied.

After stimulation he would pull his hands out from inside the apparatus and point with the left to the fabric felt by the left and with the right to the fabric felt by the right.

Surprisingly, although his left hand (right hemisphere) responded correctly, his left hemisphere vocalized: "Thats wrong!" Repeatedly he reached over with his right hand and tried to force his left extremity to point to the fabric experienced by the right (although the left hand responded correctly! His left hemisphere didn't know this, however). His left hand refused to be moved and physically resisted being forced to point at anything different. In one instance a physical

Cosmology of Consciousness

struggle ensued, the right grappling with the left with the two halves of the body hitting and scratching at each other!

Moreover, while 2-C was performing this (and other tasks), his left hemisphere made statements such as: "I hate this hand" or "This is so frustrating" and would strike his left hand with his right or punch his left arm. In these instances there could be little doubt that his right hemisphere mental system was behaving with purposeful intent and understanding, whereas his left hemisphere mental system had absolutely no comprehension of why his left hand (right hemisphere) was behaving in this manner (Joseph, 1988b).

These conflicts are not limited to behavior, TV programs, choice of clothing, or food, but to actual feelings, including love and romance. For example, the right and left hemisphere of a male split-brain patient had completely different feelings about an ex-girlfriend. When he was asked if he wanted to see her again, he said "Yes." But at the same time, his left hand turned thumbs down!

Another split-brain patient suffered conflicts about his desire to smoke. Although is left hemisphere mental system enjoyed cigarettes, his left hand would not allow him to smoke, and would pluck lit cigarettes from his mouth or right hand and put them out. He had been trying to quit for years.

Because each head contains multiple minds, similar conflicts also plague those who have not undergone split-brain surgery. Each half of the brain and thus each mental system may have different attitudes, goals and interests. As noted above, 2-C experienced conflicts when attempting to eat, watch TV, or go for walks, his right and left hemisphere mental systems apparently enjoying different TV programs or types of food (Joseph 1988b). Conflicts of a similar nature plague us all. Split-brain patients are not the first to choke on self-hate or to harm or hate those they profess to love.

Each half of the brain is concerned with different types of information, and may react, interpret and process the same external experience differently and even reach different conclusions (Joseph 1988a,b; Sperry, 1966). Moreover, even when the goals are the same, the two halves of the brain may produce and attempt to act on different strategies.

Each mental system has its own reality. Singularity of mind, is an illusion.

9. Dissociation and Self-Consciousness

The multiplicity of mind is not limited to the neocortex but includes old cortical structures, such as the limbic system (Joseph 1992). Moreover, limbic nuclei such as the amygdala and hippocampus interact with neocortical tissues creating yet additional mental systems, such as those which rely on memory and which contribute to self-reflection, personal identity, and even self-consciousness (Joseph, 1992, 1998, 1999b, 2001).

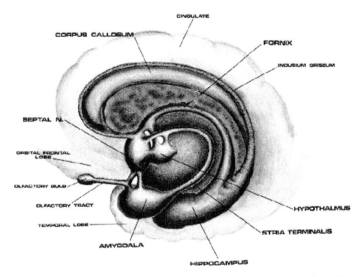

For example, both the amygdala and the hippocampus are implicated in the storage of long term memories, and both nuclei enable individuals to visualize and remember themselves engaged in various acts, as if viewing their behavior and actions from afar. Thus, you might see yourself and remember yourself engage in some activity, from a perspective outside yourself, as if you are an external witness; and this is a common feature of self-reflection and self-memory and is made possible by the hippocampus and overlying temporal lobe (Joseph, 1996, 2001).

The hippocampus in fact contains "place neurons" which cognitive map one's position and the location of various objects within the environment (Nadel, 1991; O'Keefe, 1976; Wilson & McNaughton, 1993). Further, if the subject moves about in that environment, entire populations of these place cells will fire. Moreover, some cells are responsive to the movements of other people in that environment and will fire as that person is observed to move around to different locations or corners of the room (Nadel, 1991; O'Keefe, 1976; Wilson and McNaughton, 1993).

Electrode stimulation, or other forms of heightened activity within the hippocampus and overlying temporal lobe can also cause a person to see themselves, in real time, as if their conscious mind is floating on the ceiling staring down at their body (Joseph, 1998, 1999b, 2001). During the course of electrode stimulation and seizure activity originating in the temporal lobe or hippocampus, patients may report that they have left their bodies and are hovering upon the ceiling staring down at themselves (Daly, 1958; Penfield, 1952; Penfield & Perot 1963; Williams, 1956). That is, their consciousness and sense of personal identity appears to split off from their body, such that they experience themselves as a consciousness that is conscious of itself as a conscious that is detached from the body which is being observed.

One female patient claimed that she not only would float above her body, but would sometimes drift outside and even enter into the homes of her neighbors. Penfield and Perot (1963) describe several patients who during a temporal lobe seizure, or neurosurgical temporal lobe stimulation, claimed they split-off from their body and could see themselves down below. One woman stated: "it was though I were two persons, one watching, and the other having this happen to me." According to Penfield (1952), "it was as though the patient were attending a familiar play and was both the actor and audience."

Under conditions of extreme trauma, stress and fear, the amygdala, hippocampus and temporal lobe become exceedingly active (Joseph, 1998, 1999b). Under these conditions many will experience a "splitting of consciousness" and have the sensation they have left their body and are hovering beside or above themselves, or even that they floated away (Courtois, 2009; Grinker & Spiegel, 1945; Noyes & Kletti, 1977; van der Kolk 1987). That is, out-of-body dissociative experiences appear to be due to fear induced hippocampus (and amygdala) hyperactivation.

Likewise, during episodes of severe traumatic stress personal consciousness may be fragmented and patients may dissociate and experience themselves as splitting off and floating away from their body, passively observing all that is occurring (Courtois, 1995; Grinker & Spiegel, 1945; Joseph, 1999d; Noyes & Kletti, 1977; Southard, 1919; Summit, 1983; van der Kolk 1987).

Noyes and Kletti (1977) described several individuals who experienced terror, believed they were about to die, and then suffered an out-of body dissociative experience: "I had a clear image of myself... as though watching it on a television screen." "The next thing I knew I wasn't in the truck anymore; I was looking down from 50 to 100 feet in the air." "I had a sensation of floating. It was almost like stepping out of reality. I seemed to step out of this world."

One individual, after losing control of his Mustang convertible while during over 100 miles per hour on a rain soaked freeway, reported that "time seemed to slow down and then... part of my mind was a few feet outside the car zooming above it and then beside it and behind it and in front of it, looking at and analyzing the respective positions of my spinning Mustang and the cars surrounding me. Simultaneously I was inside trying to steer and control it in accordance with the multiple perspectives I was given by that part of my mind that was outside. It was like my mind split and one consciousness was inside the car, while the other was zooming all around outside and giving me visual feedback that enabled me to avoid hitting anyone or destroying my Mustang."

Numerous individuals from adults to children, from those born blind and deaf, have also reported experiencing a dissociative consciousness after profound injury causing near death (Eadie 1992; Rawling 1978; Ring 1980). Consider for example, the case of Army Specialist J. C. Bayne of the 196th Light Infantry Brigade. Bayne was "killed" in Chu Lai, Vietnam, in 1966, after being simultaneously machine gunned and struck by a mortar. According to Bayne, when he opened his eyes he was floating in the air, looking down on his burnt and bloody body: "I could see me... it was like looking at a manikin laying there... I was burnt up and there was blood all over the place... I could see the Vietcong. I could see the guy pull my boots off. I could see the rest of them picking up various things... I was like a spectator... It was about four or five in the afternoon when our own troops came. I could hear and see them approaching... I looked dead... they put me in a bag... transferred me to a truck and then to the morgue. And from that point, it was the embalming process. I was on that table and a guy was telling jokes about those USO girls... all I had on was bloody undershorts... he placed my leg out and made a slight incision and stopped... he checked my pulse and heartbeat again and I could see that too... It was about that point I just lost track of what was taking place.... [until much later] when the chaplain was in there saying everything was going to be all right.... I was no longer outside. I was part of it at this point" (reported in Wilson, 1987, pp 113-114; and Sabom, 1982, pp 81-82).

Therefore, be it secondary to the fear of dying, or depth electrode stimulation, these experiences all appear to be due to a mental system which enables a the conscious mind to detach completely from the body in order to make the body an object of consciousness (Joseph, 1998, 1999b, 2001).

10. Quantum Consciousness

It could be said that consciousness is consciousness of something other than consciousness. Consciousness and knowledge of an object, such as a chair, are also distinct. Consciousness is not the chair. The chair is not consciousness. The chair is an object of consciousness, and thus become discontinuous from the quantum state.

Cosmology of Consciousness

Consciousness is consciousness of something and is conscious of not being that object that is conscious of. By knowing what it isn't, consciousness may know what it is not, which helps define what it is. This consciousness of not being the object can be considered the "collapse function" which results in discontinuity within the continuum.

Further, it could be said that consciousness of consciousness, that is, self-consciousness, also imparts a duality, a separation, into the fabric of the quantum continuum. Therefore this consciousness that is the object of consciousness, becomes an abstraction, and may create a collapse function in the continuum.

However, in instances of dissociation, this consciousness is conscious of itself as a consciousness that is floating above its body; a body which contains the brain. The dissociative consciousness is not dissociated from itself as a consciousness, but only from its body. That is, there is an awareness of itself as a consciousness that is floating above the body, and this awareness is simultaneously one with that consciousness, as there is no separation, no abstractions, and no objectification. It is a singularity that is without form, without dimension, without shape.

Moreover, because this dissociated consciousness appears to be continuous with itself, there is no "collapse function" except in regard to the body which is viewed as an object of perception.

Therefore, in these instances we can not say that consciousness has split into a duality of observer and observed or knower and known, except in regard to the body. Dissociated consciousness is conscious of itself as consciousness; it is self-aware without separation and without reflecting. It is knowing and known, simultaneously.

In fact, many patients report that in the dissociated state they achieve or nearly achieved a state of pure knowing (Joseph, 1996, 2001).

A patient described by Williams (1956) claimed she was lifted up out of her body, and experienced a very pleasant sensation of elation and the feeling that she was "just about to find out knowledge no one else shares, something to do with the link between life and death." Another patient reported that upon leaving her body she not only saw herself down below, but was taken to a special place "of vast proportions, and I felt as if I was in another world" (Williams, 1956).

Other patients suffering from temporal lobe seizures or upon direct electrical activation have noted that things suddenly became "crystal clear" or that they had a feeling of clairvoyance, of having the ultimate truth revealed to them, of having achieved a sense of greater awareness and cosmic clarity such that sounds, smells and visual objects seemed to have a greater meaning and sensibility, and that they were achieving a cosmic awareness of the hidden knowledge of all things (Joseph, 1996, 2001).

Although consciousness and the object of consciousness, that which is known, are not traditionally thought of as being one and the same, in dissociative states, consciousness and knowing, may become one and the same. The suggestion is of some type of cosmic unity, particularly as these patients also often report a progressive loss of the sense of individuality, as if they are merging into something greater than themselves, including a becoming one with all the knowledge of the universe; a singularity with god.

Commonly those who experience traumatic dissociative conscious states, not only float above the body, but report that they gradually felt they were losing all sense of individuality as they became embraced by a brilliant magnificent whiteness that extended out in all directions into eternity.

Therefore, rather than the dissociated consciousness acting outside the quantum state, it appears this mental state may represent an increasing submersion back into the continuity that is the quantum state, disappearing back into the continuum of singularity and oneness that is the quantum universe.

Dissociated consciousness may be but the last preamble before achieving the unity that is quantum consciousness, the unity of all things.

11. Conclusion: Quantum Consciousness and the Multiplicity of Mind

What is "Objective Reality" when the mind is a multiplicity which is capable of splitting up, observing itself, becoming blind to itself, and becoming blind to the features of the world which then cease to exist for the remaining mental realms?

Each mental system has its own "reality." Each observer is a multiplicity that engages in numerous simultaneous acts of observation. Therefore, non-local properties which do not have an objective existence independent of the "act of observation" by one mental system, may achieve existence when observed by another mental system. The "known" and the "unknown" can exist simultaneously and interchangeably and this may explain why we don't experience any macroscopic non-local quantum weirdness in our daily lives.

This means that quantum laws may apply to everything, from atoms to monkeys and woman and man and the multiplicity of mind. However, because of this

Cosmology of Consciousness

multiplicity, this could lead to seemingly contradictory predictions and uncertainty when measuring macroscopically objective systems which are superimposed on microscopic quantum systems. Indeed, this same principle applies to the multiplicity of mind, where dominant parallel streams of conscious awareness may be superimposed on other mental systems, and which may be beset by uncertainty.

Because of the multiplicity of mind, as exemplified by dissociative states, the observer can also be observed, and thus, the observer is not really external to the quantum state as is required by the standard collapse formulation of quantum mechanics. This raises the possibility that the collapse formulation can be used to describe the universe as a whole which includes observers observing themselves being observed.

The multiplicity of mind also explains why an object being measured by one mental system therefore becomes bundled up into a state where it either determinately has or determinately does not have the property being measured. Measurements performed by one mental system are not being performed by others, such that the same object can have an initial state and a post-measurement state and a final state simultaneously as represented in multiple minds in parallel, or separate states as represented by each mental realm individually.

The collapse dynamics of observation supposedly guarantees that a system will either determinately have or determinately not have a particular property. However, because the observer is observing with multiple mental systems the object can both have and not have specific properties when it is being measure and not measured, simply because it is being measured and not measured, or rather, observed and not observed or its different features observed simultaneously by multiple mental systems. Therefore, it can be continuous and discontinuous in parallel, and different properties can be known and not known in parallel simultaneously.

And the same rules apply to the mental systems which exist in multiplicity within the head of a single observer. Mental systems can become continuous or discontinuous, and can be known and not known simultaneously, in parallel. Thus, the standard collapse formulation can be used to describe systems that contain observers, as the mind/observer can be simultaneously internal and external to the described system.

The mind is a multiplicity and there is no such thing as a "single observer state." Therefore, each element may be observed by multiple observer states which perceive multiple object systems thereby giving the illusion that the object has been transformed during the collapse function. What this also implies is that contrary to the standard or Copenhagen interpretations, states may have both definite position and definite momentum at the same time.

Each mental system perceives a different physical world giving rise to multiple worlds and multiple realities which may be subordinated by one or another more dominant stream of conscious awareness.

Moreover, as the multiplicity of mind can also detach and become discontinuous with the body, whereas dissociative consciousness is continuous with itself, this indicates that the mind is also capable of becoming one with the continuum, and can achieve singularity so that universe and mind become one.

References

Akelaitis, A. J. (1945). Studies on the corpus callosum. American Journal of Psychiatry, 101, 594-599.

Bogen, J. (1979). The other side of the brain. Bulletin of the Los Angeles Neurological Society. 34, 135-162.

Bohr, N. (1934/1987), Atomic Theory and the Description of Nature, reprinted as The Philosophical Writings of Niels Bohr, Vol. I, Woodbridge: Ox Bow Press.

Bohr, N. (1958/1987), Essays 1932-1957 on Atomic Physics and Human Knowledge, reprinted as The Philosophical Writings of Niels Bohr, Vol. II, Woodbridge: Ox Bow Press.

Bohr, N. (1963/1987), Essays 1958-1962 on Atomic Physics and Human Knowledge, reprinted as The Philosophical Writings of Niels Bohr, Vol. III, Woodbridge: Ox Bow Press.

Casagrande, V. A. & Joseph, R. (1978). Effects of monocular deprivation on geniculostriate connections in prosimian primates. Anatomical Record, 190, 359.

Casagrande, V. A. & Joseph, R. (1980). Morphological effects of monocular deprivation and recovery on the dorsal lateral geniculate nucleus in prosimian primates. Journal of Comparative Neurology, 194, 413-426.

Courtois, C. A. (2009). Treating Complex Traumatic Stress Disorders: An Evidence-Based Guide Daly, D. (1958). Ictal affect. American Journal of Psychiatry, 115, 97-108.

DeWitt, B. S., (1971). The Many-Universes Interpretation of Quantum Mechanics, in B. D.'Espagnat (ed.), Foundations of Quantum Mechanics, New York: Academic Press. pp. 167–218.

DeWitt, B. S. and Graham, N., editors (1973). The Many-Worlds Interpretation of Quantum Mechanics. Princeton University Press, Princeton, New-Jersey.

Eadie, B. J. (1993) Embraced by the light. New York, Bantam

Einstein, A. (1905a). Does the Inertia of a Body Depend upon its Energy Content? Annalen der Physik 18, 639-641.

Einstein, A. (1905b). Concerning an Heuristic Point of View Toward the Emission and Transformation of Light. Annalen der Physik 17, 132-148.

Einstein, A. (1905c). On the Electrodynamics of Moving Bodies. Annalen der Physik 17, 891-921.

Einstein, A. (1926). Letter to Max Born. The Born-Einstein Letters (translated by Irene Born) Walker and Company, New York.

Gallagher, R. E., & Joseph, R. (1982). Non-linguistic knowledge, hemispheric laterality, and the conservation of inequivalance. Journal of General Psychology, 107, 31-40.

Geschwind. N. (1981). The perverseness of the right hemisphere. Behavioral Brain Research, 4, 106-107.

Grinker, R. R., & Spiegel, J. P. (1945). Men Under Stress. McGraw-Hill.

Heisenberg. W. (1930), Physikalische Prinzipien der Quantentheorie (Leipzig: Hirzel). English translation The Physical Principles of Quantum Theory, University of Chicago Press.

Heisenberg, W. (1955). The Development of the Interpretation of the Quantum Theory, in W. Pauli (ed), Niels Bohr and the Development of Physics, 35, London: Pergamon pp. 12-29.

Heisenberg, W. (1958), Physics and Philosophy: The Revolution in Modern Science, London: Goerge Allen & Unwin.

Joseph, R. (1982). The Neuropsychology of Development. Hemispheric Laterality, Limbic Language, the Origin of Thought. Journal of Clinical Psychology, 44 4-33.

Joseph, R. (1986). Reversal of language and emotion in a corpus callosotomy patient. Journal of Neurology, Neurosurgery, & Psychiatry, 49, 628-634.

Joseph, R. (1986). Confabulation and delusional denial: Frontal lobe and lateralized influences. Journal of Clinical Psychology, 42, 845-860.

Joseph, R. (1988) The Right Cerebral Hemisphere: Emotion, Music, Visual-Spatial Skills, Body Image, Dreams, and Awareness. Journal of Clinical Psychology, 44, 630-673.

Joseph, R. (1988). Dual mental functioning in a split-brain patient. Journal of Clinical Psychology, 44, 770-779.

Joseph, R. (1992) The Limbic System: Emotion, Laterality, and Unconscious Mind. The Psychoanalytic Review, 79, 405-455.

Joseph, R. (1996). Neuropsychiatry, Neuropsychology, Clinical Neuroscience, 2nd Edition. Williams & Wilkins, Baltimore.

Joseph, R. (1998). Traumatic amnesia, repression, and hippocampal injury due to corticosteroid and enkephalin secretion. Child Psychiatry and Human Development. 29, 169-186.

Joseph, R. (1999a). Frontal lobe psychopathology: Mania, depression, aphasia, confabulation, catatonia, perseveration, obsessive compulsions, schizophrenia. Psychiatry, 62, 138-172.

Joseph, R. (1999b). The neurology of traumatic "dissociative" amnesia. Commentary and literature review. Child Abuse & Neglect. 23, 71-80.

Joseph, R. (2000). Limbic language/language axis theory of speech. Behavioral and Brain Sciences. 23, 439-441.

Joseph, R. (2001). The Limbic System and the Soul: Evolution and the Neuroanatomy of Religious Experience. Zygon, the Journal of Religion &

Science, 36, 105-136.

Joseph, R., & Casagrande, V. A. (1978). Visual field defects and morphological changes resulting from monocular deprivation in primates. Proceedings of the Society for Neuroscience, 4, 1978, 2021.

Joseph, R. & Casagrande, V.A. (1980). Visual field defects and recovery following lid closure in a prosimian primate. Behavioral Brain Research, 1, 150-178.

Joseph, R., Forrest, N., Fiducia, N., Como, P., & Siegel, J. (1981). Electrophysiological and behavioral correlates of arousal. Physiological Psychology, 1981, 9, 90-95.

Joseph, R., Gallagher, R., E., Holloway, J., & Kahn, J. (1984). Two brains, one child: Interhemispheric transfer and confabulation in children aged 4, 7, 10. Cortex, 20, 317-331.

Joseph, R., & Gallagher, R. E. (1985). Interhemispheric transfer and the completion of reversible operations in non-conserving children. Journal of Clinical Psychology, 41, 796-800.

Nadel, L. (1991). The hippocampus and space revisited. Hippocampus, 1, 221-229.

Neumann, J. von, (1937/2001), "Quantum Mechanics of Infinite Systems. Institute for Advanced Study; John von Neumann Archive, Library of Congress, Washington, D.C.

Neumann, J. von, (1938), On Infinite Direct Products, Compositio Mathematica 6: 1-77.

Neumann, J. von, (1955), Mathematical Foundations of Quantum Mechanics, Princeton, NJ: Princeton University Press.

Noyes, R., & Kletti, R. (1977). Depersonalization in response to life threatening danger. Comprehensive Psychiatry, 18, 375-384.

O'Keefe, J. (1976). Place units in the hippocampus. Experimental Neurology, 51-78-100.

Pais, A. (1979). Einstein and the quantum theory, Reviews of Modern Physics, 51, 863-914.

Penfield, W. (1952) Memory Mechanisms. Archives of Neurology and Psychiatry, 67, 178-191.

Penfield, W., & Perot, P. (1963) The brains record of auditory and visual experience. Brain, 86, 595-695.

Rawlins, M. (1978). Beyond Death's Door. Sheldon Press.

Ring, K. (1980). Life at Death: A Scientific Investigation of the Near-Death Experience. New York: Quill.

Sabom, M. (1982). Recollections of Death. New York: Harper & Row.

Southard, E. E. (1919). Shell-shock and other Neuropsychiatric Problems. Boston.

Sperry, R. (1966). Brain bisection and the neurology of consciousness. In F. O. Schmitt and F. G. Worden (eds). The Neurosciences. MIT press.

van der Kolk, B. A. (1985). Psychological Trauma. American Psychiatric Press.

Williams, D. (1956). The structure of emotions reflected in epileptic experiences. Brain, 79, 29-67.

Wilson, I. (1987). The After Death Experience. Morrow.

Wilson, M. A., & McNaughton, B. L. (1993). Dynamics of the hippocampus ensemble for space. Science, 261, 1055-1058.

Cosmology of Consciousness

Logic of Quantum Mechanics and Phenomenon of Consciousness

Michael B. Mensky

P.N. Lebedev Physical Institute, Russian Academy of Sciences, Moscow,

Abstract

The phenomenon of consciousness, including its mystical features, is explained on the basis of quantum mechanics in the Everett's form. Everett's interpretation (EI) of quantum mechanics is in fact the only one that correctly describes quantum reality: any state of the quantum world is objectively a superposition of its classical counterparts, or "classical projections". They are "classically incompatible", but are considered to be equally real, or coexisting. We shall call them alternative classical realities or simply classical alternatives. According to the Everett's interpretation, quantum reality is presented by the whole set of alternative classical realities (alternatives). However, these alternatives are perceived by humans separately, independently from each other, resulting in the subjective illusion that only a single classical alternative exists. The ability to separate the classical alternatives is the main feature of what is called consciousness. According to the author's Extended Everett's Concept (EEC), this feature is taken to be a definition of consciousness (more precisely, consciousness as such, not as the complex of processes in the conscious state of mind). It immediately follows from this definition that turning the consciousness off (in sleeping, trance or meditation) allows one to acquire access to all alternatives. The resulting information gives rise to unexpected insights including scientific insights.

1 INTRODUCTION

Strange as this may seem, we do not know what is the nature ofconsciousness, especially of the very strange features of consciousness which resemble mystical phenomena. The most familiar examples of these mysterious phenomena are scientific insights (of course, we mean only "great insights" experienced by great scientists). Some people supposed that the mystery of consciousness may be puzzled out on the basis of quantum mechanics, the science which is mysterious itself.

This viewpoint has been suggested, as early as in 20th years of 20th century, by the great physicist Wolfgang Pauli in collaboration with the great psychologist Carl Gustav Jung. They supposed particularly that quantum mechanics may help to explain strange psychic phenomena observed by Jung and called "synchronisms". Jung told of a synchronism if a series of the events happened such that these events were conceptually close but their simultaneous (synchronous) emergence could not be justified causally. For example he observed causally unjustified, seemingly accidental, appearance of the image of fish six times during a single day. The work of Pauli and Jung on this topic was not properly published and was later completely forgotten, but it became popular in last decades (see about this in Enz, 2009).

The idea of connecting consciousness with quantum mechanics was suggested by some other authors, mostly without referring Pauli and Jung. In the last three decades this idea was supported by Roger Penrose (Penrose, 1991), (Penrose, 1994). He particularly remarked that people manage to solve such problems which in principle cannot be solved with the help of computers because no algorithms exist for their solving. Penrose suggested that quantum phenomena should be essential for explaining the work of brain and consciousness.

Usually attempts to explain consciousness on the base of quantum mechanics follow the line of consideration that is natural for physicists. Everything must be explained by natural sciences, may be with accounting quantum laws. Therefore, in order to explain consciousness, one has to apply quantum mechanics for analysis of the work of brain. For example, the work of brain may be explained as the work of quantum computer instead of classical one. Thus, it is usually assumed, explicitly or implicitly, that consciousness must be derived from the analysis of the processes in brain.

The approach proposed by the author in 2000 does not include this assumption. This approach is based on the analysis of the logical structure of quantum mechanics, and the phenomenon of consciousness is derived from this purely logical analysis rater than from the processes in brain. The actual origin of the concept of consciousness is, according to this approach, specific features of the concept of reality accepted in quantum mechanics (contrary to classical physics) and often called quantum reality.

Quantum reality has its adequate presentation in the so-called Everett's interpretation (EI) of quantum mechanics known also under name of many-worlds interpretation (Everett III, 1957). The approach of the author is based on the Everett's form of quantum mechanics and called Extended Everett's Concept (EEC).

Some physicists, having in mind purely classical concept of reality, consider the Everett's interpretation of quantum mechanics too complicated and "exotic". However, it is now experimentally proved that reality in our world is quantum, and the conclusions based on classical concept of reality, are not reliable. The comprehension of the concept of quantum reality was achieved after long intellectual efforts of genius scientists. Unfortunately, the ideas of Pauli and Jung were not properly estimated and timely used. The first whose thoughts about quantum reality became widely know was Einstein who, in the work with his coauthors, suggested so-called Einstein-Podolski-Rosen paradox (Einstein et al., 1935). Much later John Bell formulated now widely known Bell's theorem (Bell, 1964), (Bell, 1987) which provided an adequate tool for direct quantum-mechanical verification of the concept of quantum reality, the Bell's inequality. Less than in 20 years the group of Aspect experimentally proved (Aspect et al., 1981) that the Bell's inequality is violated in some quantum processes, and therefore reality in our world is quantum.

Most simple and convenient formulation of quantum reality was given even earlier that the Bell's works, in 1957, by the Everett's interpretation of quantum mechanics. It was enthusiastically accepted by the great physicists John Archibald Wheeler and Brice Dewitt, but was not recognized by the wide physical community until last decades of 20th century, when the corresponding intellectual base was already prepared. From this time the number of adepts of the Everett's interpretation grows permanently. Results of this difficult but very important process of conceptual clarification of quantum mechanics justify the appreciation of the Everett's interpretation as the only correct form of quantum mechanics. It is exciting that, as an additional prize, the Everett's interpretation explains the mysterious phenomenon of consciousness.

Quantum mechanics in the Everett's form implies coexisting "parallel worlds", or parallel classical realities. This clearly expresses the difference of quantum reality from the common classical reality. According to the author's Extended Everett's Concept (Menskii, 2000), consciousness is the ability to perceive the Everett's parallel worlds separately, independently from each other. An immediate consequence of this assumption is that the state of being unconscious makes all parallel realities available without the separation. This leads to irrational insights and other "mystical" phenomena. This is the central point of the theory making it plausible. Indeed, it is well known (from all spiritual schools and from deep psychological researches) that the strange abilities of consciousness arise just in the states of mind that are close to being unconsciousness (sleeping, trance or meditation).

Cosmology of Consciousness

2 PARALLEL WORLDS

According to the Everett's "many-worlds" interpretation of quantum mechanics, quantum mechanics implies coexistence of "parallel classical worlds", or alternative classical realities. This follows from the arguments (see details below) including the following points:

• The very important specific feature of quantum systems is that their states are vectors. This means that a state of any quantum system may be a sum (called also superposition) of other states of the same system. All the states which are the counterparts of this sum, are equally real, i.e. they may be said to coexist. This is experimentally proved for the states of microscopic systems (such as elementary particles or atoms).

• However, this should be valid also for macroscopic systems (consisting of many atoms). It follows from the logical analysis of the measurements of microscopic systems. Indeed, let a microscopic system S be measured with the help of a macroscopic measuring device M. If the state of S is a sum of a series of states, then, after the measurement, the state of the combined system (S and M) is also a sum (each its term consisting of a state of S and the corresponding state of M).

• Different states of the measuring device, by the very definition of a measuring device, have to be macroscopically distinct. Therefore, a macroscopic system may be in the state which is a sum (superposition) consisting of the states which are incompatible (alternative to each other) from the point of view of classical physics. However, quantum mechanics requires them to "coexist" (in the form of a sum, or superposition).

• Therefore, classically incompatible states of our world (alternative classical realities) must coexist as a sort of "parallel worlds" which are called Everett's worlds.

Let us now give some details of these arguments. What does it mean that the states of a quantum system are vectors? If vpi are states of some quantum system, then $\psi = \psi 1 + \psi 2 + \bullet \bullet \bullet$ is also a state of the same system, called a superposition of the states ψi (counterparts of the superposition).

This feature is experimentally proved for microscopic systems (such as elementary particles or atoms), but it has to be valid also for macroscopic systems. This follows from the analysis of measurements with microscopic measured systems and macroscopic measuring devices.

The conclusion following from this analysis is that, even for a macroscopic system, its state ψ may be a superposition of other states of this system, which have evident classical interpretation (are close to some classical states of the system) while the state ψ has no such an interpretation:

$$\psi = \psi 1 + \psi 2 +...+ \psi n... (1)$$

It is important that the states ψ here may be macroscopically distinct, therefore, from classical point of view incompatible, alternative to each other, presenting alternative classical realities. Nevertheless, it follows from quantum theory that even such macroscopically distinct states ψi may be in superposition, i.e. may coexist. Let us formulate the above situation a little bit more precisely. The quantum system is in the state denoted by the state vector and only this state objectively exists (however, taken as a whole, has no classical interpretation). The counterparts $\wedge$ of the superposition are in fact classical projections of the objectively existing quantum state ψ. These classical projections describe images of the quantum system which arise in consciousness of an observer (therefore, they concern the subjective aspect of quantum reality). This status of the classical projections will be made more transparent below.

In the following we shall use ψ for the state of our (quantum) world as a whole. The components ψi of the superposition will be alternative classical states of this world (more precisely, quasiclassical, i.e. the states as close to classical as is possible for the quantum world).

In the Everett's interpretation of quantum mechanics the states ψ are called Everett's worlds. We shall use also the terms "parallel worlds", "alternative classical realities", "classical alternatives" or simply "alternatives". In case if such alternatives are superposed (as in Eq. (1)), we shall say that they "coexist". Of course this word is nothing else than a convenient slang, meaning in fact "to form a superposition" or "to be in superposition". The status of the "coexistence" is connected with the consciousness and subjective perception of the world, which will be explained below.

3 EXTENDED EVERETT'S CONCEPT (EEC)

We see that the alternative classical realities in our quantum world may coexist (as components of a superposition presenting the state of the quantum world). Subjectively however each observer perceives only a single "classical alternative". These two assertions seem to contradict to each other. Are they in fact compatible? We shall show how this seeming contradiction is resolved in the Everett's interpretation (EI) and how it may be taken as a basis for the theory of consciousness and the unconscious if the EI is properly extended.

3.1 Everett's "Many-Worlds" Interpretation One may naively think that the picture of the world arising in his consciousness (the picture of a single classical alternative) is just what objectively exists. However, EI of quantum mechanics (unavoidably following from the logics of quantum mechanics applied to the phenomenon of quantum measurements) claims that it is only the superposition of all alternatives (as in Eq. (1)) that objectively exists.

The seemingly strange and counter-intuitive presentation of objective reality (in the Everett's form of quantum mechanics) as the set of many objectively coexisting classical realities adequately expresses quantum character of reality in our (quantum) world.

The single alternatives (components of the superposition) present various subjective perceptions of this quantum reality in an observer's consciousness. The natural question arises how the multiplicity of the "classical pictures" that may arise in our consciousness may be compatible with the subjective sensory evidence of only a single such picture.

This is the most difficult point of the EI and the reason why this interpretation has not been readily accepted by the physicists. This point may be made more transparent if it is presented in the terms of "Everett's worlds" as it has been suggested by Brice DeWitt.

Thus, all of the separate components of the superposition (classical alternatives, or Everett's worlds) are declared by Everett to be "equally real". No single alternative may be considered to be the only real, while the others being potentially possible but not actualized variants of reality (this might be accepted in classical physics, but not in quantum mechanics, because of the special character of quantum reality).

To make understanding of the EI easier, Brice DeWitt proposed (De-Witt & Graham, 1973) to think that each observer is present in each of the Everett's worlds. To make this even more transparent, one may think that a sort of twins ("clones") of each observer are present in all Everett's worlds. Subjective perception is the perception of a single twin. Objectively the twins of the given observer exist in all Everett's worlds, each of them perceiving corresponding alternative pictures.

Each of us subjectively perceives around him a single classical reality. However, objectively the twins, or clones, of each of us perceive all the rest realities. It is important that all twins of the given observer are equally real. It is incorrect to say that there is "I" and there are my twins, which are not "I". All twins embody me as an observer, each of them can be called "I".

Thus, the concept of "Everett's world" allows to make the Everett's presentation of quantum reality more transparent. We prefer to verbalize the same situation in another way (Menskii, 2000). We shall say that all alternative classical projections of the quantum world's state objectively exist, but these projections are separated in consciousness. Subjective perception of the quantum world by human's consciousness embraces all these classical pictures, but each picture is perceived independently of the rest.

Regardless of the way of wording, the Everett's assumption of objectively coexisting classical alternatives implies that all these alternatives may be accessible for an observer in some way or another. Yet it is not clear how the access to "other alternatives" (different from one subjectively perceived) may be achieved for the given observer. It is usually claimed that the EI does not allow observing "other alternatives". This makes the interpretation non-falsifiable and thus decreases its value.

It turns out however that the EI may be improved in such a way that that the question "How one can access to other alternatives?" is answered in a very simple and natural way. This improvement is realized in the author's Extended Everett's Concept. The (improved) EI becomes then falsifiable, although in a special sense of the word (see below Sect. 3.2).

3.2 Extension of the Everett's Interpretation Starting with the above mentioned formulation (that classical alternatives are separated in consciousness), the present author proposed (Menskii, 2000) to accept a stronger statement: consciousness is nothing else than separation of the alternatives.

This seemingly very small step resulted in important consequences. Finally the so-called Extended Everett's Concept (EEC) has been developed (Menskii, 2005), (Menskii, 2007), (Mensky, 2005, 2007, 2010).

The first advantage of EEC is that the logical structure of the quantum theory is simplified as compared with the EI.

The point is that the formulation "alternatives are separated in consciousness" (accepted in one of the possible formulations of EI) includes two primary (not definable) concepts, "consciousness" and "alternative separation". These concepts have no good definitions. One may object that many different definitions have been proposed for the notion "consciousness". This is right, but all these definitions concern in fact mental and sensual processes in brain rather than "consciousness as such", while the latter (more fundamental) notion has no good definition.

After the notions "consciousness" (more precisely, "consciousness as such") and "separation of alternatives" are identified (as it is suggested in EEC), only one of these concepts remains in the theory. Therefore, EEC includes only one primary concept instead of two such concepts in the EI. The logical structure of the theory is simplified after its extension.

Much more important is that EEC gives a transparent indication as to how an observer may obtain access to "other alternatives" (different from the alternative subjectively perceived by him). This very important question remains unanswered (or answered negatively) in the original form of the EI. This is seen from the following argument.

If consciousness = separation (of the alternatives from each other), then absence of consciousness = absence of separation.

Therefore, turning off consciousness (in sleeping, trance or meditation) opens access to all classical alternatives put together, without separation between them. Of course, the access is realized then not in the form of visual, acoustic or other conscious images or thoughts. Nothing at all can be said about the form of this access. However, if we accept EEC, then we may definitely conclude that the access is possible in the unconscious state of mind.

This of course has very important consequences. The access to the enormous "data base" consisting of all alternative classical realities enables one to acquire valuable information, or rather knowledge. This information is unique in the sense that it is unavailable in the conscious state when only a single alternative is subjectively accessible. One may suppose that a part of this unique information may be kept on returning to the usual conscious state of mind and recognized in the form of usual conscious images and thoughts.

Thus, when going over to the unconscious state, one obtains the information, or knowledge, which is in principle unavailable in the usual conscious state.

This information is unique first of all because it is taken from "other" classical alternatives (different from one subjectively observed). There is however something more that makes this information unique and highly valuable. All alternatives together form a representation of the quantum state of the world (vector $\wedge$ in Eq. (1)). Time evolution of this state vector, according to quantum laws, is reversible. This means that, given at some time moment, this vector is known also at all other times. Therefore, information about "all alternatives together" (i.e. about the state vector of the world) includes information from any time moments in future and past. This information may be thought of as being obtained with the help of a "virtual time machine".

Evidently, this makes "irrational" inspirations (including scientific insights) possible. Here is a simple example. Let a scientist be confronted with a scientific problem and consider a number of hypotheses for solving this problem. Going, by means of the above mentioned virtual time machine, into future and backward, the scientist may find out what of these hypotheses will be confirmed by future experiments or proved with the help of the future theories. Then, on returning to the conscious state, he will unexpectedly and without any rational grounds get certain about which of these hypotheses has to be chosen.

Remark that it is not necessary, for making use of such a virtual time machine, to turn off consciousness completely (although it is known that some important discoveries were made in sleeping, or rather after awakening). It is enough to disconnect it from the problem under consideration. This is why solutions of hard problems are sometimes found not during the work on these problems but rather during relaxation.

Preliminary "rational" work on the problem is however necessary. The deep investigation of all data concerning the problem enables the consciousness to form a sort of query (clear formulation of the problem and its connections with all relevant areas of knowledge). Then the query will be worked out in the unconscious state (during relaxation) and will result in an unexpected insight.

It is clear that not only scientific problems can be solved in this way, but also problems of general character. Quite probable that, besides ordinary intuitive guesses, we meet in our experience examples of super-intuitive insights of the type described above. Anybody knows that many efficient solutions come in the morning just after awakening. This fact may be an indirect confirmation of the ability of super-intuition.

4 PROBABILISTIC MIRACLES

Thus, Extended Everett's Concept (EEC) leads to the conclusion that unconscious state of mind allows one to take information "from other alternatives" that reveals itself as unexpected insights, or direct vision of truth.

Another consequence is feasibility of even more weird phenomenon looking as arbitrary choice of reality. Let us describe this ability in a special case of what can be called "probabilistic miracles".

Consider an observer who subjectively perceives one of the alternative classical realities at the present time moment t0. Let in a certain future moment $t > t0$ some event E may happen, but with a very small probability p. Call it the objective probability of the event E and suppose that this probability is small.

According to the Everett's interpretation of quantum mechanics, at time t two classes of alternative classical realities will exist so that the event E happens in each alternative of the first class and does not happen in the alternatives of the second class. The twins of our observer will be objectively present in each of the alternatives (this is the feature of Everett's worlds, see Sect. 3.1). However, subjectively our observer will feel to be in one of them. With some probability p it will turn out to be the alternative of the first class. The probability p may be called subjective probability of the event E for the given observer.

It is accepted in the Everett's interpretation that subjective and objective probabilities coincide, $p = p$. However, in the context of EEC we may assume that they may differ and, moreover, the observer may influence the value of subjective probability p'. Let us assume that the observer prefers the event E to happen. Then he can enlarge the subjective probability of this event, i.e. the probability to find himself subjectively at time t just in that classical reality in which this rare event actually happens.

Thus, besides the objective probability of any event, there is a subjective probability of this event for the given observer, and the observer may in principle influence the subjective probability. In the above mentioned situation, an event under consideration can happen according to usual laws of the natural sciences, but with small probability. This means that the objective probability of this event is small, it may seem even negligible. It is important that the objective probability is non-zero. One may say that among all alternatives at the moment t few of the alternatives correspond to the pictures of the world in which the event happens, and much more alternatives correspond to the pictures of the world where the event does not happen.

However, according to EEC, an observer can, simply by the force of his consciousness, make the subjective probability of this event close to unity. Then very likely he will find himself at the moment t in one of those classical realities where the event does happen.

The subjective experience of such an observer will evidence that the objectively improbable event may be realized by the effort of his will. This looks like a miracle. However, this is a miracle of a special type, which may be called "probabilistic miracle".

Probabilistic miracles essentially differ from "absolute" miracles that happen in fairy tales. The difference is that the event realized as a probabilistic miracle (i.e. "by the force of consciousness") may in principle happen in a quite natural way, although with a very small probability. This small but nonzero probability is very important. Particularly, because of fundamental character of probabilistic predictions in quantum mechanics, it is in principle impossible to prove or disprove the unnatural (miraculous) character of the happening.

Indeed, if the objectively rare event happens, the person who has strongly desired for it to happen is inclined to consider the happening as a result of his will. Yet any skeptic may insist in this situation that the event occurred in a quite natural way: what happened, was only a rare coincidence. The secret is in the nature of the concept of probability: if the probability of some stochastic event is equal p, then in a long series consisting of N tests the event will happen pN times (very rarely for small p). But it is in principle impossible to predict in which of these tests the event will happen. Particularly, it may happen even in the very first test from the long series of them.

The latter is a very interesting and general feature of the phenomena "in the area of consciousness" as they are treated in EEC. These phenomena in principle cannot be unambiguously assigned to the sphere of natural events (obeying the laws of natural sciences) or to the sphere of spiritual or psychic phenomena (which are treated by the humanities and spiritual doctrines). Impossibility to do this may be called relativity of objectiveness.

Synchronisms studied by Jung may be considered to be probabilistic miracles. One who observes a subject or event which somehow attracts his special attention, involuntarily thinks about it (often even not clearly fixing his thoughts). According to the above said, he may increase the subjective probability of immediately observing something similar or logically connected.

Some Biblical miracles can also be explained as probabilistic miracles. An example is the miracle at the Sea of Galilee where Jesus calmed the raging storm (Matthew 8:23-27). This was completely natural event. Wonderful was only the fact that the storm ceased precisely at that moment when this was necessary for Jesus and his disciples. The probability of "timely" occurring this natural event was of course very small. The miracle was probabilistic.

5 CONCLUDING REMARKS

We shortly followed in this paper the main ideas of Extended Everett's Concept (EEC) about nature of consciousness. Let us briefly comment on the further development of this theory.

All that has been discussed above, makes sense for humans (possessing consciousness) and may be partly for higher animals. However, the theory may be generalized to give the quantum concept of life in a more general aspect. Thus modified theory is meaningful not only for humans, but for all living beings (belonging to the type of life characteristic for Earth). The idea of the generalization is follows (see (Mensky, 2010) for details).

The main point of EEC is the identification of the "separation of classical alternatives" with the human's consciousness. Now we have to identify this quantum concept with the ability of the living beings to "perceive the quantum world classically". This is an evident generalization of consciousness but for all living beings. Instead of "consciousness" in EEC we have now "classical perceiving of quantum reality" which means that the alternative classical realities (forming the state of the quantum world) are perceived separately from each other.

Similarly to what we told about consciousness (in case of humans), this ability of living beings to "classically perceive the quantum world" is necessary for the very phenomenon of life (of local type). The reason is the same: elaborating efficient strategy of surviving is possible only in a classical world which is "locally predictable". Existing objectively in the quantum world, any creature is living in each of the classical realities separately from all the rest classical realities. Life is developing parallely in the Everett's parallel worlds.

Remark by the way that from this point of view "existing" and "living" are different concepts. Important difference is that existing (evolution in time) of the inanimate matter is determined by reasons while living of the living beings is partly determined also by goals (first of all the goal of survival). Let us make some other remarks concerning philosophical or rather meta-scientific aspects of EEC.

This theory shows that a conceptual bridge between the material (described by natural sciences) and the ideal (treated by the humanities and spiritual doctrines) does exist.

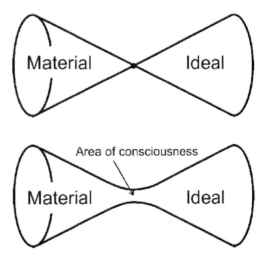

Fig 1: Carl Jung illustrated the relation between sphere of the material and sphere of the ideal as two cones with common vertex (above). In EEC the "area of consciousness" is common for both spheres (below). Some of the phenomena in the area of consciousness cannot be unambiguously assigned to exclusively one of the spheres: objectiveness is relative.

This bridge is determined in EEC in a concrete way, but the idea of such a bridge is not novel (see Fig. 1). The creators of quantum mechanics from the very beginning needed the notion of the "observer's consciousness" to analyze conceptual problems of this theory (the "problem of measurement"). In fact, the difficulties in solving these problems were caused by the insistent desire to construct quantum mechanics as a purely objective theory. Nowadays it becomes clear that there is no purely objective quantum theory. Objectiveness is relative (see Sect. 4).

There is a very interesting technical point in relations between the material and the ideal. We see from the preceding consideration that the description of the ideal, or psychic (consciousness and the unconscious), arises in the interior of quantum mechanics when we consider the whole world as a quantum system. This provides the absolute quantum coherence which is necessary for the conclusions that are derived from EEC. Usually only restricted systems are considered in quantum mechanics. The resulting theory is purely material. Ideal (psychic) elements arise as the specific aspects of the whole world. The unrestricted character of the world as a quantum system is essential for this (cf. the notion of microcosm).

All these issues demonstrate the specific features of the present stage of quantum theory. Including theory of consciousness (and the unconscious) in the realm of quantum mechanics (starting by Pauli and Jung and now close to being accomplished) marks a qualitatively new level of understanding quantum mechanics itself. The present stage of this theory can be estimated as the second quantum revolution. When being completed, it will accomplish the intellectual and philosophical revolution that accompanied creating quantum mechanics in the first third of 20th century.

Cosmology of Consciousness

References

Aspect, A., Grangier, P., Roger, G. (1981). Phys. Rev. Lett., 47, 460.

Bell, J. S. (1964). Physics, 1, 195. Reprinted in (Bell, 1987).

Bell, J. S. (1987). Speakable and Unspeakable in Quantum Mechanics. Cambridge Univ. Press, Cambridge.

DeWitt, B. S., Graham, N. (Eds.) (1973). The Many-Worlds Interpretation of Quantum Mechanics. Princeton Univ. Press, Princeton, NJ.

Einstein, A., Podolsky, B., Rosen, N. (1935). Phys. Rev., 47, 777.

Enz, C. P. (2009). Of Matter and Spirit. World Scientific Publishing Co., New Jersey etc. Everett III, H. (1957). Rev. Mod. Phys., 29, 454.

Menskii, M. B. (2000). Quantum mechanics: New experiments, new applications and new formulations of old questions. Physics- Uspekhi, 43, 585-600.

Menskii, M. B. (2005). Concept of consciousness in the context of quantum mechanics. Physics-Uspekhi, 48, 389-409.

Menskii, M. B. (2007). Quantum measurements, the phenomenon of life, and time arrow: Three great problems of physics (in Ginzburg's terminology). Physics-Uspekhi, 50, 397-407.

Mensky, M. (2007). Postcorrection and mathematical model of life in Extended Everett's Concept. NeuroQuantology, 5, 363-376. www.neuroquantology.com, arxiv:physics.gen-ph/0712.3609.

Mensky, M. (2010). Consciousness and Quantum Mechanics: Life in Parallel Worlds (Miracles of Consciousness from Quantum Mechanics). World Scientific Publishing Co., Singapore.

Mensky, M. B. (2005). Human and Quantum World (Weirdness of the Quantum World and the Miracle of Consciousness). Vek-2 publishers, Fryazino. In Russian.

Penrose, R. (1991). The Emperor's New Mind: Concepting Computers, Minds, and the Laws of Physics. Penguin Books, New York.

Penrose, R. (1994). Shadows of the Mind: a Search for the Missing Science of Consciousness. Oxford Univ. Press, Oxford.

Evolution of Paleolithic Cosmology and Spiritual Consciousness, and the Temporal and Frontal Lobes

Rhawn Joseph, Ph.D.
Emeritus, Brain Research Laboratory, California

Abstract

Complex mortuary rituals and belief in the transmigration of the soul, of a world beyond the grave, has been a human characteristic for at least 100,000 years. The emergence of spiritual consciousness and its symbolism, is directly linked to the evolution of the temporal and frontal lobes and to the Neanderthal and Cro-Magnon peoples, and then the first cosmologies, 20,000 to 30,000 years ago. These ancient peoples of the Upper and Middle Paleolithic were capable of experiencing love, fear, and mystical awe, and they carefully buried those they loved and lost. They believed in spirits and ghosts which dwelled in a heavenly land of dreams, and interned their dead in sleeping positions and with tools, ornaments and flowers. By 30,000 years ago, and with the expansion of the frontal lobes, they created symbolic rituals to help them understand and gain control over the spiritual realms, and created signs and symbols which could generate feelings of awe regardless of time or culture. Because they believed souls ascended to the heavens, the people of the Paleolithic searched the heavens for signs, and between 30,000 to 20,000 years ago, they observed and symbolically depicted the association between woman's menstrual cycle and the moon, patterns formed by stars, and the relationship between Earth, the sun, and the four seasons. These include depictions of 1) the "cross" which is an ancient symbol of the fours seasons and the Winter/Summer solstice and Spring/Fall equinox; 2) the constellations of Virgo, Taurus, Orion/Osiris, the Pleiades, and the star Sirius; 3) and the 13 new moons in a solar year. Although it is impossible to date these discoveries with precision, it can be concluded that spiritual consciousness first began to evolve over 100,000 years ago, and this gave birth to the first heavenly cosmologies over 20,000 years ago.

1. THE TRANSMIGRATION OF THE SOUL

Belief in the transmigration of the soul, of a life after death, of a world beyond the grave, has been a human characteristic for at least 100,000 years, as ancient graves and mortuary rites attest (e.g., Belfer-Cohen & Hovers, 1992; Butzer, 1982; McCown, 1937; Rightmire, 1984; Schwarcz et al., 1988; Smirnov, 1989; Trinkaus 1986). Even ancient "archaic" humans, despite their small brains and primitive intellectual, linguistic, cognitive, and mental capabilities, and who wondered the planet over 120,000 years ago, carefully buried their dead (Butzer, 1982; Joseph 2000a; Rightmire, 1984). And like modern Homo sapiens, they prepared the recently departed for the journey to the Great Beyond: across the sea of dreams, to the land of the dead, the heavens, the realm of the ancestors and the gods.

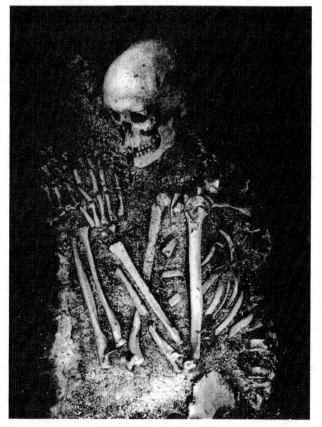

Figure: Paleolithic burial in sleeping position.

Throughout the Middle and Upper Paleolithic it was not uncommon for tools and hunting implements to be placed beside the body, even 100,000 years ago, for the dead would need them in the next world (Belfer-Cohen & Hovers, 1992; McCown, 1937; Trinkaus, 1986). A hunter in life he was to be a hunter in death, for the ethereal world of the Paleolithic was populated by spirits and souls of bear, wolf, deer, bison, and mammoth (e.g., Campbell, 1988; Joseph 2001, 2002; Kuhn, 1955). Moreover, food and water might be set near the head in case the spirit hungered or experienced thirst on its long sojourn to the heavenly Hereafter. And finally, fragrant blossoming flowers and red ocher might be sprinkled upon the bodies (Solecki, 1971) along with the tears of those who loved them.

Given the relative paucity of cognitive, cultural, and intellectual development among Middle Paleolithic Neanderthal and "archaic" humans, and the likelihood that they had not yet acquired modern human speech (Joseph, 1996, 2000b), evidence of spiritual concerns among these peoples demonstrates the great antiquity of belief in an after-life and the soul. Humans began evolving a spiritual consciousness over 100,000 years ago. Seventy thousand years later, this consciousness would give birth to the first cosmologies.

2. MIDDLE PALEOLITHIC SPIRITUALITY

When humans first became aware of a "god" or "gods" cannot be determined. Nevertheless, the antiquity of religious and spiritual belief extends backwards in time to over 100,000 years. It is well established that Neanderthals and other Homo Sapiens of the Middle Paleolithic (e.g. 150,000 to 35,000 B.P.) and Upper Paleolithic (35,000 B.P. to 10,000 B.P.) engaged in complex ritualistic behavior. These rituals are evident from the manner in which they decorated their caves and the symbolism associated with death (Akazawa & Muhesen 2002; Conrad & Richter 2011; Harvati & Harrison 2010; Kurten 1976; Mellars 1996).

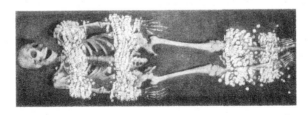

Figure: Cro-Magnon burial.

Neanderthals (a people who lived in Europe, Russia, Iraq, Africa, and China from around 150,000 to 30,000 B.P.), have been buried in sleeping positions with the body flexed and lying on its side. Some were laid to rest with limestone blocks placed beneath the head like a pillow—as if they were not truly dead but merely asleep (Akazawa & Muhesen 2002; Harvati & Harrison 2010).

Sleep and dreams have long been associated with the spirit world, and it is through dreams that gods including the Lord God worshipped by Jews, Christians, and Muslims, are believed to have communicated their thoughts, warnings, intentions, and commands. Throughout the ages (Campbell 1988; Freud, 1900; Jung 1945, 1964), and as repeatedly stated in the Old Testament and the Koran, dreams have been commonly thought to be the primary medium in which gods and human interact (Joseph 2002). Insofar as the ancients (and many moderns folks) were concerned, dreams served as a doorway, a portal of entry to the spirit world through which "God," His angels, or myriad demons could make their intentions known.

Figure: Paleolithic burial in sleeping position.

It is through dreams that one is able to come into direct contact with the spirit world and a reality so magical and profoundly different yet as real as anything experienced during waking. It is through dreams that ancient humans came to believe the spiritual world sits at the boundaries of the physical, where day turns to dusk, the hinterland of the imagination where dreams flourish and grow. And it is while dreaming that one's own soul may transcend the body, to soar like an eagle, or to commune with the spirits of loved ones who reside in heaven along side the gods.

Neanderthals prepared their dead for this great and final journey, by laying their loved ones to rest so that they would sleep with the spirits and dream of heavenly eternity.

Neanderthals have also been buried surrounded by goat horns placed in a circle, with reindeer vertebrae, animal skins, stone tools, red ocher, and in one grave, seven different types of flowers (Solecki 1971). In one cave (unearthed after 60,000 years had passed), a deep chamber was discovered which housed a single skull which was surrounded by a ring of stones (Harvati & Harrison 2010; Mellars 1996). Moreover, Neanderthals buried bears at a number of sites including Regourdou. At Drachenloch they buried stone "cysts" containing bear skulls (Kurten 1976); hence, "the clan of the cave bear."

Figure: Neanderthal burial. This Neanderthal was buried with 7 different types of flowers.

Yet others were buried with large bovine bones above the head, with limestone blocks placed on top of the head and shoulders, and with heads severed coupled with evidence of ritual decapitation, facial bone removal, and cannibalism (Belfer-Cohen and Hovers 1992; Binford 1968; Harold 1980; Smirnov 1989; Solecki 1971). In one site, dated to over 100,000 years B.P., Neanderthals decapitated eleven of their fellow Neanderthals, and smashed their faces beyond recognition.

It therefore seems apparent that Neanderthals not only engaged in complex rituals, but they believed in spirits, ghosts, and a life after death. Hence the sleeping position, stone pillows, stone tools and food. They were preparing the departing spirit for the journey to the Hereafter and the land of dreams. However, they were also incapacitating their enemies, even after death, to prevent these souls from terrorizing the living or their dreams

3. SPIRITS, SOULS, GHOSTS & THE LAND OF DREAMS

The fact that so many of the Neanderthal dead were buried in a sleeping position implies an association between sleep and dreams. Since all vertebrates so far studied demonstrate REM (dream) sleep, it can be assumed that Neanderthals dreamed. Among ancient (and even modern peoples) it was believed that souls and spirits could wonder about while people sleep and dream (Brandon 1967; Frazier 1950; Harris 1993; Jung 1945, 1964; Malinowkski 1990). Some believed the soul could escape the body via the mouth or nostrils while dreaming and that the spirit could leave the body and engage in various purposeful acts or interact with other souls including the soul or spirit of those who had died. The spirit and soul were believed to hover about in human-like, ghostly vestiges, at the fringes of reality, the hinterland where day turns into night (Campbell 1988; Frazier 1950; Jung 1964; Malinowski 1954; Wilson 1951); and it is at night when people dream.

Figure: Neanderthal burial.

And as is the case with modern day humans, it can be assumed the ancients, including Neanderthals, had dreams by which they were transported or exposed to a world of magic and untold wonders which obeyed its own laws of time, space, and motion. It is through dreams that humans came to believe the spiritual world sits at the boundaries of the physical, where day turns to dusk, the hinterland of the mind where imagination and dreams flourish and grow (Frazier, 1950; Jung, 1945, 1964; Malinowkski, 1954); hence the tendency to bury the dead in a sleeping position even 100,000 years ago.

It is also via dreams that humans came to know that spirits and lost souls populated the night. The dream was real and so too were the ghosts, gods and demons who thundered and condemned and the phantoms that hovered at the edge of night. Although but a dream, like modern humans, our ancient ancestors experienced this through the senses, much as the physical world is experienced. Dreams were real and they were taken seriously. Moreover, during dreams, both the living and the dead may be encountered. Thus, we can surmise that Neanderthals had similar dreams and may have dreamed about ghosts and wondering spirits,

It is also appears that they feared the dead, and were terrified by the prospect that certain souls might haunt the living. They were afraid of ghosts, and frightened by the possibility that just as one might awake from sleep after visiting the land of the dead, the dead might also awake from this deathlike slumber. The dead, or at least their personal souls, had to be prevented from causing mischief among the living; especially dreaded enemies who had been killed. Hence, the ritual decapitation, facial bone removal, the smashing of faces, the removal of arms, hands, and legs, and placement of heavy stones upon the body.

It can be concluded, therefore, that almost 100,000 years ago, primitive humans had already come to believe in ghosts, souls, spirits, and a continuation of "life" after death. And, they also took precautions, in some cases, to prevent certain spirits and souls from being released from a dead body and returning to cause mischief among the living, which is why, in the case or powerful enemies, the Neanderthals would cut off heads and hands. They went to great lengths to obliterate all aspects of that dreaded individual's personal identity; e.g., smashing the face beyond recognition.

Of course, the fact that these Neanderthals were buried does not necessarily imply that they held a belief in "God." Rather, what the evidence demonstrates is that Neanderthals were capable of very intense emotions and feelings ranging from murderous rage to love to spiritual and superstitious awe. Although no god is implied, Neanderthals held spiritual and mystical beliefs involving the transmigration of the soul and all the horrors, fears, and hopes that accompany such feelings and beliefs. Although the Neanderthals had not discovered god, they stood upon the threshold.

4. THE NEANDERTHALS: A CHARACTER STUDY

Neanderthals were short, brutish, and an exceedingly violent, murderous people, as the remnants of their skeletons preserved for so many eons attests. Many of their fossils still betray the cruel ravages of deliberately and violently inflicted wounds (Conrad & Richter 2011; Harvati & Harrison 2010; Mellars 1996).

They also appear to have systematically engaged in female infanticide, and displayed a willingness to eat almost anything on four or two legs—including other Neanderthals. In one site, dated to over 100,000 years B.P., Neanderthals decapitated eleven of their fellow Neanderthals, and then enlarged the base of each skull (the foramen magnum) so the brains could be scooped out and presumably eaten. Even the skulls of children were treated in this fashion.

In fact, they would throw the bones and carcasses of other Neanderthals into the refuse pile. In one cave, a collection of over 20 Neanderthals were found mixed up with the remains of other animals and garbage. Presumably, these were enemies or just hapless strangers, innocent cave dwellers who were attacked and sometimes eaten after being brutally killed.

Figure: (Left) Two well preserved crania from northern European male Neanderthals. Reproduced from M. H. Wolpoff (1980), Paleo-Anthropology. New York, Knopf. (Right) Neanderthal Male.

Hence, with the obvious exception of "friends," mates, and family, Neanderthals often saw one another as a potential meal, and had almost no regard for a stranger's innate humanness. These were a violent, murderous, ritualistic people, and strangers were often brutally killed and eaten.

These characteristics are also associated with religious fervor. Among ancient and present day peoples, violence, murder, ritual cannibalism, and the sacrifice of children are common religious practices. The Five Books of Moses, are replete with stories of the mass murder and the genocide of non-Jews who were seen as subhuman, including pregnant women and children.

"...when you approach a town, you shall lay seizure to it, and when the Lord your god delivers it into your hand, you shall put all its males to the sword.... In the towns of the people which the Lord your god is giving you as a heritage, you shall not let a soul remain alive." Exodus 20:15-18; Deuteronomy 20:12-16.

"When Israel had killed all the inhabitants of Ai....and all of them, to the last man had fallen by the sword, all the Israelites turned back to Ai and put it to the sword...until all the inhabitants of Ai had been exterminated... and the king of Ai was impaled on a stake and it was left lying at the entrance to the city gate." Deuteronomy 8:24-29.

I polluted them with their own offerings, making them sacrifice all their first-born, which was to punish them, so that they would learn that I am Yahweh (Ezekiel 20:25-36. See also Ezekiel 22:28-29). "This very day you defile yourselves in the presentation of your gifts by making your children pass through the fire of all your fetishes." (Ezekiel 20:31). "A blessing on him who seizes your babies and dashes them against rocks." (Psalm 137:9).

Figure: Aztec and Indian natives were burnt alive in groups of 13 to honor Jesus Christ and his 12 disciples.

As is well known, the Spanish and Catholic missionaries, acting at the behest of the Catholic Popes (and their Spanish/Catholic Sovereigns), continued these genocidal practices once they invaded the Americas during the 1500's and up through the 19th century. As the Catholic Dominican Bishop Bartolom de Las Casas reported to the Pope: the Aztec and Indian natives were hung and burnt alive "in groups of 13... thus honoring our Savior and the 12 apostles."

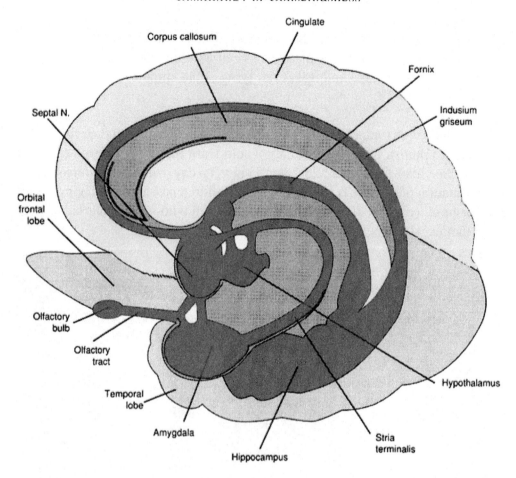

5. The Limbic System: Love, Violence, & Spirituality.

Violence and murder are also under the control of the limbic system, the amygdala in particular (Joseph 1992, 1994, 1996, 1998). And it is the limbic system which mediates religious and spiritual experience and which provides much of the emotion and imagery which appears in dreams (Joseph 1988, 2000a, 2001, 2002).

However, it is also the limbic system which subserves feelings of love and affection, and the ability to form long-term attachments. Thus we see that Neanderthals provided loving care for friends and family who had been injured or maimed, enabling them to live many more years despite their grievous injuries. For example, the skeleton of one Neanderthal male, who was about age 45 when he died, had been nursed for a number of years following profoundly crippling injuries. His right arm had atrophied, and his lower arm and hand had apparently been ripped or bitten off, and his left eye socket, right shoulder, collarbone, and both legs were badly injured. Obviously someone loved and tenderly cared for this man. He was no doubt a father, a husband, a brother, and son, and someone in his family not only provided long term loving care to make him comfortable in this life, but prepared him for the next life as well (Mellars 1996).

The ability to feel love is a function of the limbic system, the amygdala in particular which is buried in the depths of the temporal lobe.

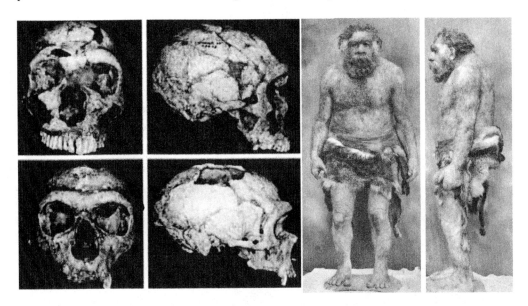

6. THE NEANDERTHAL BRAIN & TEMPORAL LOBES

An examination of Neanderthal skulls and endocasts of the inner skull provides a gross indication of the size and configuration of their brains. Based on physical indices, the temporal lobe was well developed and little different from that of modern humans.

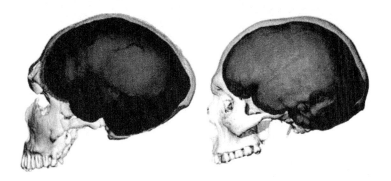

Figure: (Left)Neanderthal. (Right) Modern human.

Likewise, given the great antiquity of the limbic system, it can be surmised that the Paleolithic human limbic system was well developed, and similar to the limbic system of modern humans (Joseph 2000a, 2001; 2002).

The Neanderthals were not a very intelligent or tidy people and were unable to fashion complex tools which, along with other indices, suggests they were unable think complex thoughts. Yet they were people of passion who experienced profound emotions and love; made possible by the limbic system and temporal lobe. In fact, it is because they had the limbic capacity to experience love, spiritual awe, and religious concerns, that these expressions of love continued following death of those they loved, as it has been conclusively demonstrated that these brain structures mediate these functions (Joseph 1992, 1994, 1998a, 2001, 2002). Thus the Neanderthals carefully buried their dead, providing them with food and even sprinkling the bodies with seven different types of blooming, blossoming, fragrant flowers (Belfer-Cohen & Hovers, 1992; McCown, 1937; Solecki 1971; Trinkaus, 1986).

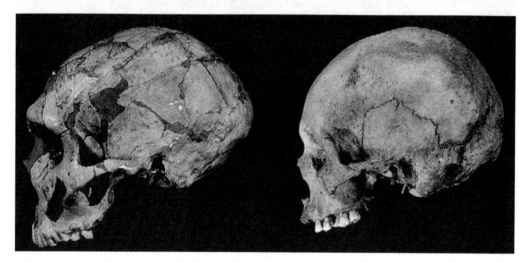

Figure: (Left)Neanderthal. (Right) Modern human.

In overall size, the posterior portion of the Neanderthal brain, i.e. the occipital and superior parietal lobes, were slightly larger in length and breadth than the modern human brain on average (Joseph 1996, 2000b; Wolpoff, 1980); a reflection of the environment in which they dwelled and the neural capacities their life style required-- the body moving in visual space as Neanderthals spent most of their non-sleeping hours searching for food. The occipital and superior parietal areas are directly concerned with visual analysis and positioning the body in space (Joseph 1986, 1996). As male and female Neanderthals spent a considerable amount of their time engaged in hunting activities, scanning the environment for prey and running and throwing in visual space were more or less ongoing concerns. A large occipital and superior parietal lobe would reflect these activities.

By contrast, concerns about the dead, and attendant mortuary rituals are activities linked to the temporal lobes. The temporal lobes and underlying limbic structures (amygdala, hippocampus), could be likened the seat of the soul and the senior executive of the personality. It is the temporal lobes and the amygdala and hippocampus which have been directly implicated in the generation of religious feelings and supernatural experiences including visions of floating above the body, seeing angels and devils, and what has been described as the after-death and near-death experiences (Joseph 1996, 1998b, 1999a,b, 2000a, 2001, 2002).

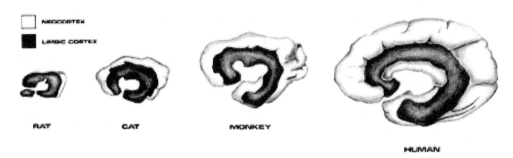

The amygdala (which is buried in the depths of the anterior temporal lobe) enables us to hear "sweet sounds," recall "bitter memories," or determine if something is spiritually significant, sexually enticing, or good to eat and makes it possible to experience the spiritually sublime. It is concerned with the most basic animal emotions, and allows us to store affective experiences in memory or even to reexperience them when awake or during the course of a dream in the form of visual, auditory, or religious or spiritual imagery. The amygdala also enables an individual to experience emotions such as love and religious rapture, as well as the ecstasy associated with orgasm and the dread and terror associated with the unknown.

In fact, the amygdala (in conjunction with the hippocampus and overlying temporal lobe) contributes in large part to the production of very bizarre, unusual and fearful mental phenomenon including dissociative states, feelings of depersonalization, and hallucinogenic and dream-like recollections involving threatening men, naked women, the experience of god, as well as demons and ghosts and pigs walking upright dressed as people (Daly 1958; Gloor 1997; Halgren 1992; Horowitz et al. 1968; MacLean 1990; Penfield and Perot 1963; Schenk, and Bear 1981; Slater and Beard 1963; Subirana and Oller-Daurelia 1953; Trimble 1991; Weingarten, et al. 1977; Williams 1956). Moreover, some individuals report communing with spirits or receiving profound knowledge from the Hereafter, following amygdala stimulation or abnormal activation (Penfield and Perot 1963; Subirana and Oller-Daurelia, 1953; Williams 1956).

Intense activation of the temporal lobe, hippocampus, and amygdala has been reported to give rise to a host of sexual, religious and spiritual experiences; and chronic hyperstimulation can induce an individual to become hyper-religious or visualize and experience ghosts, demons, angels, and even "God," as well as claim demonic and angelic possession or the sensation of having left their body (Bear 1979; Gloor 1992; Horowitz et al. 1968; MacLean 1990; Penfield and Perot 1963; Schenk, and Bear 1981; Weingarten, et al. 1977; Williams 1956).

Much of the visual, emotional, and hallucinatory aspects of dream sleep, have their source in the temporal lobe and underlying limbic system structures (Joseph 1992, 1994, 1996, 1998, 2001, 2002). It is the evolution of the temporal lobes, this "transmitter to god" which also explains why even primitive humanity likely believed in spirits, souls, and ghosts, and practiced complex mortuary rites for those they feared or loved.

7. THE BIG BANG IN SYMBOLIC THINKING: THE CRO-MAGNON FRONTAL LOBES

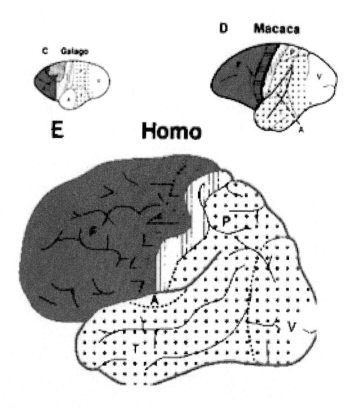

There is considerable evidence that over the course of human history, the temporal lobe evolved at a faster and earlier rate than the frontal lobe (Joseph 1996, 2000b, Gloor, 1997). Likewise, the temporal lobes mature more rapidly than the frontal lobes over the course of human ontological development (Joseph, 1982, 1996, 1998b, 1999a, 2000b,c).

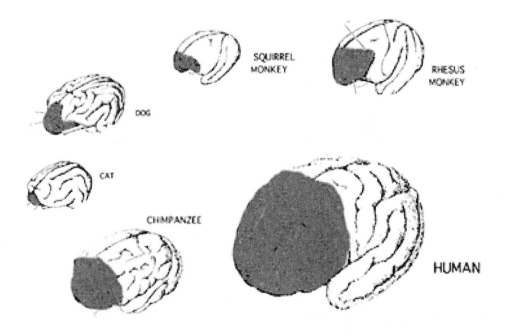

Figures: (above & previous page) Comparison of the frontal lobes in different species

The Neanderthals were blessed with a well developed temporal lobe, whereas more anterior regions of the brain, the frontal lobes, remained little different from more ancient ancestral primate species. However, with the evolution of the Cro-Magnon people, the brain mushroomed in size, with much of that development in the frontal lobes

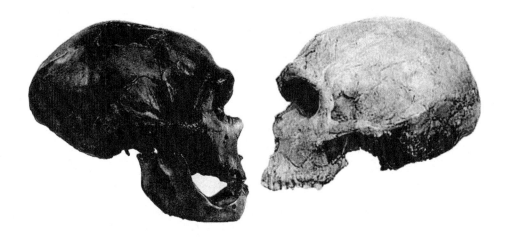

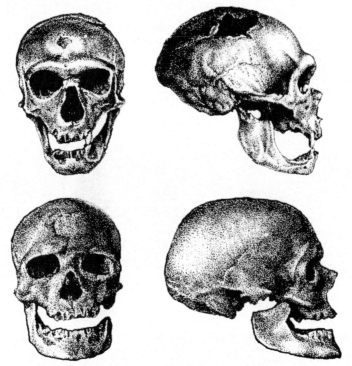

Figure: (Top) Neanderthal skull. (Bottom) Cro-Magnon Skull

As based on cranial comparisons and endocasts of the inside of the skull, and using the temporal and frontal poles as reference points, it has been demonstrated that the brain has tripled in size over the course of human evolution, and that the frontal lobes significantly expanded in length and height during the Middle to Upper Paleolithic transition (Blinkov and Glezer 1968; Joseph 1993; MacLean 1990; Tilney 1928; Weil 1929; Wolpoff 1980).

It is obvious that the height of the frontal portion of the skull is greater in the six foot tall, anatomically modern Upper Paleolithic H. sapiens (Cro-Magnon) versus Neanderthal and archaic H. sapiens (Joseph 1996, 2000b; Tilney, 1928; Wolpoff 1980). The evolution and expansion of the frontal lobe is also evident when comparing the skills and creative and technological ingenuity of the Cro-Magnons, vs the Neanderthals (Joseph 1993, 1996, 2000b).

Therefore, whereas the temporal, occipital and parietal lobes were well developed in archaic and Neanderthals, the frontal lobes would increase in size by almost a third in the transition from archaic humans to Cro-Magnon (Joseph 1996, 2000a,b, 2001).

It is the evolution of the frontal lobes which ushered in a cognitive and creative big bang which gave birth to a technological revolution and complex spiritual rituals and beliefs in shamans and goddesses and their relationship to the heavens, and thus the moon and the stars.

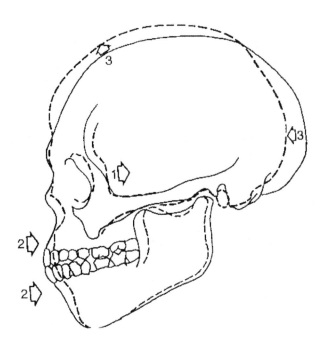

Neanderthals died out as a species around 30,000 years ago; but for at least 10,000 years they shared the planet with the Cro-Magnon people. The Cro-Magnon men stood 6ft tall on average and the males and females were very handsome and beautiful, with thin hips, aquiline noses, prominent chins, small even perfect teeth, and high rounded foreheads. There was nothing ape-like or Neanderthal about these people.

The Cro-Magnon cerebrum was also significantly larger than the Neanderthal brain, with volumes ranging from around 1600 to 1880 cc on average compared with 1,033 to 1,681 cc for Neanderthals (Blinikov & Glezer 1968; Clark 1967; Day 1996; Holloway 1985; Roginskii & Lewin 1955; Wolpoff 1980). In fact, the Cro-Magnon brain is one third larger than the modern human brain, i.e. 1800 cc vs 1350 ccs. However, a distinguishing characteristic of the Cro-Magnon brain, was the massively developed frontal lobes.

The frontal lobes are the senior executive of the brain and are responsible for initiative, goal formation, long term planning, the generation of multiple alternatives, and the consideration of multiple consequences (Joseph, 1986a, 1999a). The frontal lobes are the source of creativity, imagination, and what has been described as free will (Joseph 1996, 2011). Through interactional pathways maintained with brainstem, limbic system, thalamus and the primary receiving and association areas in the neocortex (Petrides & Pandya 1999, 2001)), the frontal lobes have access to every stage of information analysis, and are able to coordinate and regulate attention, memory, personality, and information processing throughout the brain so as to direct intellectual, creative, artistic, symbolic, and cognitive processes (Joseph, 1986a, 1999a).

It is well established that the frontal lobes enable humans to think symbolically, creatively, imaginatively, to plan for the future, to consider the consequences of certain acts, to formulate secondary goals, and to keep one goal in mind even while engaging in other tasks, so that one may remember and act on those goals at a later time (Joseph 1986, 1990b, 1996, 1999c). Selective attention, planning skills, and the ability to marshal one's intellectual resources so as to anticipate the future rather than living in the past, are capacities clearly associated with the frontal lobes.

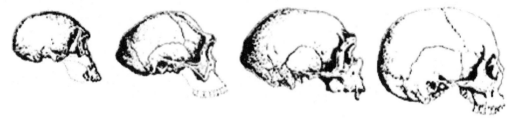

Figure: Comparison of the frontal cranium over the course of "evolution:" from Australopithecus, to H. Habilis, to H. erectus. to modern humans. Note obvious expansion of the anterior portion of the skull frontal lobe.

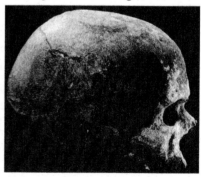

Figure: Cro-Magnon Skull

The frontal lobes are associated with the evolution of "free will" (Joseph 1986, 1996, 1999c, 2011b) and the Cro-Magnon were the first species on this planet to exercise that free will, shattering the bonds of environmental/ genetic determinism by doing what had never been done before: After they emerged upon the scene over 35,000 years ago, they created and fashioned tools, weapons, clothing, jewelry, pottery, and musical instruments that had never before been seen. They created underground Cathedrals of artistry and light, adorned with magnificent multi-colored paintings ranging from abstract impressionism to the surreal and equal to that of any modern master (Breuil, 1952; Leroi-Gourhan 1964, 1982). And they used their skills to carve the likeness of their female gods.

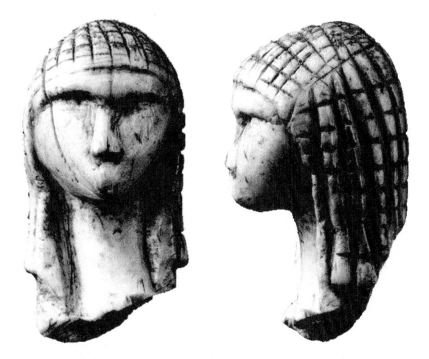

Figure: Paleolithic Goddess: Venus de Brassempouy.

Thirty five thousand years ago, Cro-Magnon were painting animals not only on walls but on ceilings, utilizing rich yellows, reds, and browns in their paintings and employing the actual shape of the cave walls so as to conform with and give life-like dimensions, including the illusion of movement to the creature they were depicting (Breuil, 1952; Leroi-Gourhan 1964, 1982). Many of their engraving on bones and stones also show a complete mastery of geometric awareness and they often used the natural contours of the cave walls, including protuberances, to create a 3-dimensional effect (Breuil, 1952; Leroi-Gourhan 1964, 1982).

The drawing or carving often became a harmonious or rather, an organic part of the object, wall, ceiling, or tool upon which it was depicted. The Cro-Magnon drew and painted scenes in which animals mated, defecated, fought, charged, and/or were fleeing and dying from wounds inflicted by hunters. The Cro-Magnon cave painters were exceedingly adept at recreating the scenes of everyday

life. Moreover, most of the animals were drawn to scale, that is, they were depicted in their actual size; and all this, 30,000 years ago (e.g. Chauvet, et al., 1997).

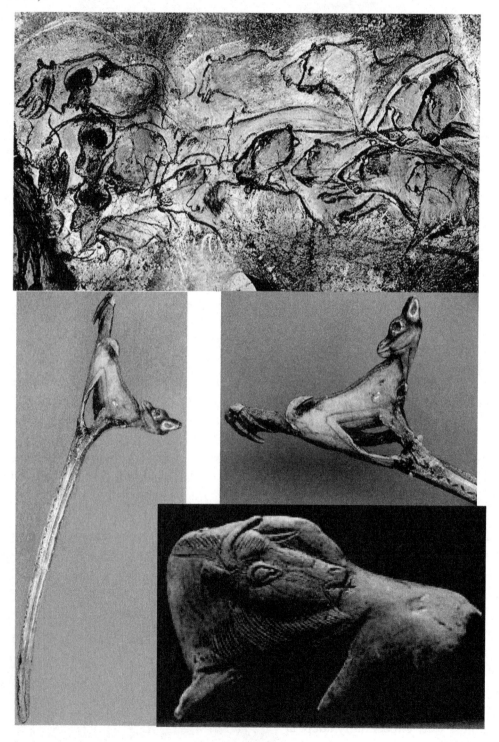

They created art that was meant to be looked at, owned and admired, and for trade, as jewelry and household decorations, and as highly prized possessions as well as for religious reverence. Picasso was awestruck by these Paleolithic masterpieces, and complained that although 30,000 years had elapsed, "we have learned nothing new. We have invented nothing."

(Left) "Leaping Horse" carved as part of a Paleolithic spear thrower. (Below). Cro-Magnon tool kits.
(Opposite Page). Carved & engraved spear throwers.

With the evolution of the Cro-Magnon people, the frontal lobes mushroomed in size and there followed an explosion in creative thought and technological innovation. The Cro-Magnon were intellectual giants. They were accomplished artists, musicians, craftsmen, sorcerers, and extremely talented hunters, fishermen, and highly efficient gatherers and herbalists. And they were the first to contemplate the heavens and the cosmos which they symbolized in art.

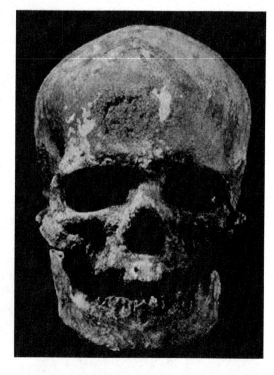

From the time of Homo Erectus (1.9 million to 500,000 B.P), humans had utilized fire to keep warm, to provide light, to cook their food, and to ward off animals. However, the Cro-Magnon learned over 30,000 years ago how to make fire using the firestone; iron pyrite which when repeatedly struck with a flint makes sparks which can easily ignite brush. They also created the first rudimentary blast furnaces which were capable of emitting enormous amounts of heat, so as to fire clay. This was accomplished by digging a tiny tunnel into the bottom of the hearth which allowed air to be drawn in. Indeed, 30,000 years ago these people were making fire hardened ceramics and clay figures of animals and females with bulging buttocks and breasts—which are presumed to be the first goddesses and fertility fetishes(Breuil, 1952; Leroi-Gourhan 1964).

Many of these female figurines were shaped so that they tapered into points so they could be stuck into the ground or into some other substance either for ornamental or supernatural purposes, e.g., household goddesses, fertility figures, and earth mothers (Breuil, 1952). In fact, much of the art produced, be it finely crafted "laurel leafs" or other artistic masterpieces, served ritual, spiritual, and esthetic functions.

It is the evolution of the Cro-Magnon and their massive frontal lobes which ushered in a cognitive and creative big bang which gave birth to complex spiritual rituals and beliefs in shamans, goddesses, and the cosmos (Joseph 2001, 2001, 2002).

By contrast, Neanderthals, archaics, and other peoples of the Middle Paleolithic were not very smart, and lived in the "here and now." They had little capacity for creative or abstract thought, and constructed and made and used the same simple stone tools over and over again for perhaps 200,000 years, until around 35,000 B.P., with little variation or consideration of alternatives (Binford, 1982; Gowlett, 1984; Mellars, 1989, 1996).

Figure: Neanderthal tools.

Cosmology of Consciousness

As neatly summed up by Hayden (1993, p. 139), "as a rule, there is no evidence of private ownership or food storage, no evidence for the use of economic resources for status or political competition.... no ornaments or other status display items, no skin garments requiring intensive labor to produce, no tools requiring high energy investments, no intensive regional exchange for rare items like sea shells or amber, no competition for labor to produce economic surpluses and no corporate art or labor intensive rituals in deep cave recesses to impress onlookers and help attract labor."

Neanderthals greatly lacked in creativity, initiative, imagination, and tended to create simple stone tools that served a single purpose. They tended to live in the immediacy of the present, with little ability to think about or consider the distant future or engage in creative or abstract thought (Binford, 1973, 1982; Dennell, 1985; Mellars, 1989, 1996). These are capacities associated with the frontal lobes.

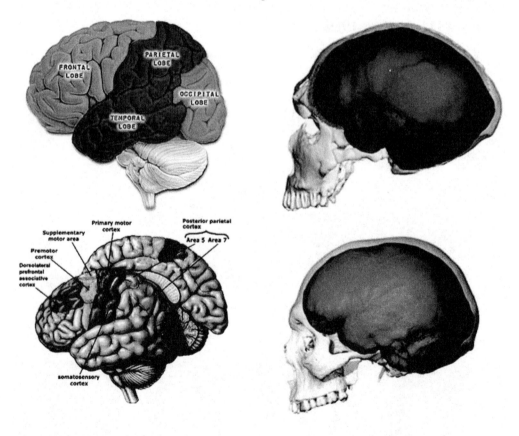

Figures: (Top Left) Modern human brain. (Upper Right) Neanderthal cranium with endocasts of their brains superimposed. (Lower) Modern human

About one third of the frontal lobe, i.e. the motor areas, are concerned with initiating, planning, and controlling the movement of the body and fine motor functioning (Joseph 1990b, 1999c). It is this part of the "archaic" and Neanderthal frontal lobe that appears to be most extensively developed. However, the more anterior frontal lobes are concerned with wholly different functions ranging from creative thought to analytical and planning skills (Joseph 1986, 1996, 1999c), and it is this region of the brain which exploded in size with the evolution of the Cro-Magnon people.

Thus, whereas mortuary rites and primitive spirituality can be associated with the temporal lobes and the Neanderthals, it was not until the evolutionary expansion of the non-motor regions of the frontal lobes that spirituality and concepts of the soul could be expressed through abstract symbolism. Therefore, the evolution of spirituality preceded abstract concepts which could be associated with religiosity; all of which in turn are directly linked to the differential evolution of the frontal and temporal lobes.

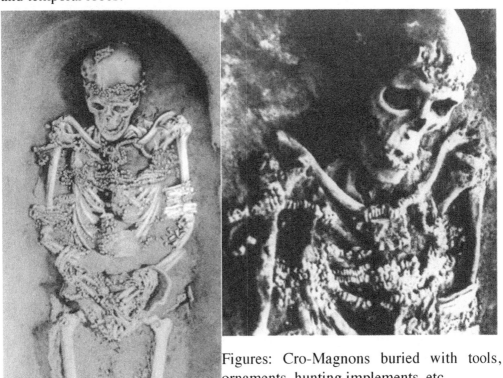

Figures: Cro-Magnons buried with tools, ornaments, hunting implements, etc.

8. CRO-MAGNON UPPER PALEOLITHIC SPIRITUAL CONSCIOUSNESS: THE BIRTH OF THE GODS

The Cro-Magnon practiced complex religious rituals and apparently were the first peoples to have arrived at the conception of "god." However, it was not a male god who they worshipped but female goddesses who were attended by animals and shaman.

Beginning over 30,000 years ago the Cro-Magnon were painting, drawing, and etching bear and mammoth, dear and horse, and pregnant females and goddesses in the recesses of dark and dusky caverns (Bandi 1961; Breuil, 1952; Chauvet et al., 1996; Leroi-Gourhan 1964, 1982; Prideaux 1973). The pregnant females include Venus statuettes, many of which were fertility goddesses.

Figure: Paleolithic fertility rites. Dancing Paleolithic Goddess surrounded by female dancers.

A pregnant woman is a symbol of fertility. However, so to is a slim, big busted female. The Cro-Magnon were able to draw both. In fact, these were the first people to paint and etch what today might be considered Paleolithic porn: slim, shapely, naked and nubile young maidens in various positions of repose.

These naked females were not drawn for the sake of prurient interests. These were fertility gods associated with the heavens and the stars.

The Cro-Magnon paid homage to a number of goddesses who was associated with the fertility of the earth, as well as the moon and the stars. One great goddess, linked to the moon was carved in limestone over the entrance to an underground cathedral in Laussel, France, perhaps 20,000 years ago. She was painted in the colors of life

and fertility: blood red. Her left hand still rests upon her pregnant belly whereas in her right hand she holds the horned crescent of the moon which is engraved with thirteen lines, the number of new moon cycles in a solar year. She was a goddess of life, linked to the mysteries of the heavens and the magical powers of the moon whose 29 day cycle likely corresponded with the Cro-Magnon menstrual cycle which issues from a woman's life-giving womb. The Cro-Magnon believed in gods. God was a woman linked to the Moon, and the earth was her womb from which life would spring anew.

Figure: A mother goddess, holding in her hand the symbol of the moon (or a bison's horn) with 13 lines, which is the number of menstrual/lunar cycles in a solar year. Her other hand rests upon her pregnant belly. This goddess was carved outside the entrance to an underground Paleolithic Cathedral, in Laussel.

Figure: A bull head shaman, the legs and vaginal area of a woman, the head/body of a lion, facing a cross, and painted on a breast-like protrusions. The shaman appears to be mating with the female.

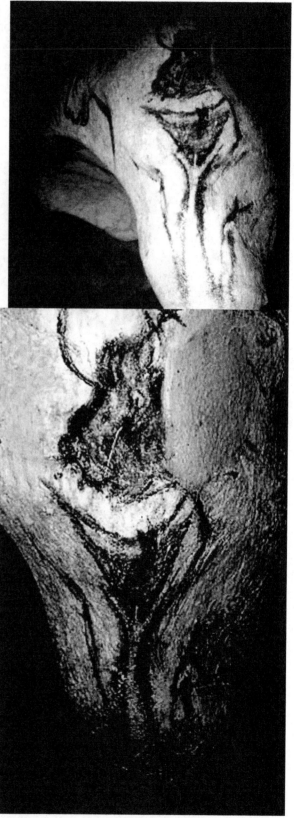

These great underground cathedrals may have also served as the Earth-womb of the goddess, where souls were reborn as men, women, and animals. Specific locations within the Earth-womb were of were of ritualistic, mystical, and spiritual significance in that many paintings were in out of the way places where one had to crawl long distances through tiny spaces and along rather tortuous routes to get to them (Leroi-Gourhan 1964). Moreover, for almost 20,000 years, subsequent generations of Cro-Magnon artists crawled to these same difficult to reach locations to repaint or paint over existing drawings which were hidden away in deep recesses of these dark underground caverns that were extremely difficult to find (Leroi-Gourhan 1964). This indicates that the location within the cave was of particular mystical and ritualistic importance. And not just the location but the journey to these hidden recesses may have been of mystical significance perhaps relating to birth, or rebirth following death.

9. SPIRITS, SOULS, AND SORCERERS

As is evident from their cave art and symbolic accomplishments, the nether world of the Cro-Magnon and other peoples of the Upper Paleolithic, was haunted by the spirits and souls of the living, the dead, and those yet to be born (Brandon 1967; Campbell 1988; Prideaux 1973).

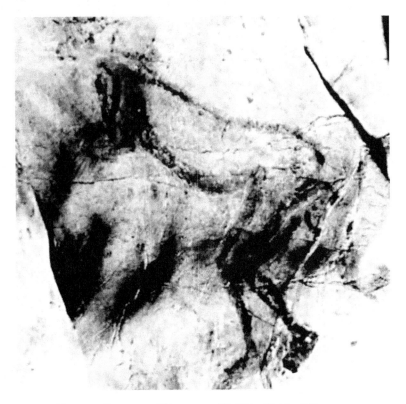

Figure: Sorcerer/Shaman: Half bull / half human

Upper Paleolithic peoples apparently believed these souls and spirits could be charmed and controlled by hunting magic, and through the spells of sorcerers and shamans. Hence, in conjunction with the worship of the goddess, the Cro-Magnons also relied on shamans dressed as animals.

Hundreds of feet beneath the earth, the likeness of one ancient shaman attired in animal skins and stag antlers, graces the upper wall directly above the entrance to the 20,000-25,000 year-old grand gallery at Les Trois-Freres in southern France (Breuil, 1952; Leroi-Gourhan 1964). Galloping, running, and swirling about this ancient sorcerer are bison, stag, horse, deer, and presumably their spirits and souls. Images of an almost identical "sorcerer" appear again in ancient Sumerian and Babylonian inscriptions fashioned four to six thousand years ago (Joseph 2000a). The "sorcerer" has a name: "Enki"-the god of the double helix.

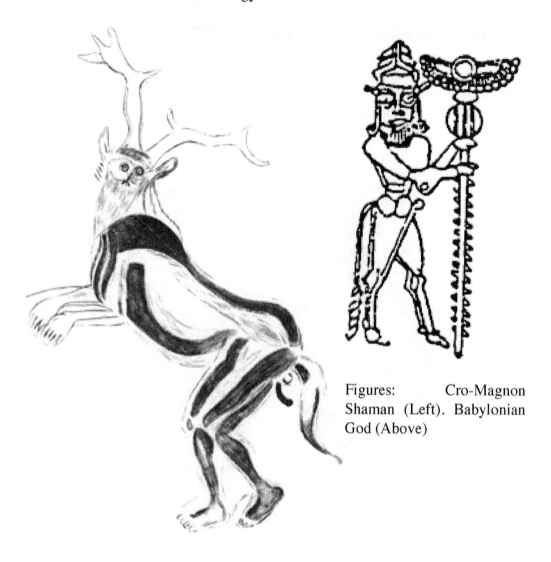

Figures: Cro-Magnon Shaman (Left). Babylonian God (Above)

10. UNDERGROUND CATHEDRALS: EMBRACED BY THE LIGHT

To view these Cro-Magnon paintings and shrines, one had to enter the hidden entrance of an underground cave, and crawl a considerable distance, sometimes hundreds of yards, through a twisting, narrowing, pitch black tunnel before reaching these Upper Paleolithic underground cathedrals which were shrouded in darkness. Here the Cro-Magnon would light candles and lamps, performing various rituals as the painted animals and spirits wavered in the cave light.

The nature and location of the Cro-Magnon cathedrals, which have been found throughout Europe, and the nature of the tortuous routes to get to them, and the effect of cave light bringing these paintings to life, is significant as it embraces features associated with after death experiences as retold by present day (as well as ancient) peoples.

Quantum Physics, Neuroscience of Mind

Figures: Lascaux cave, France.

In the ancient Egyptian and Tibetan Books of the Dead, and has been reported among many of those who have undergone a "near death" or "life after death" experience, being enveloped in a dark tunnel is commonly experienced soon after death. It is only as one ascends the tunnel that one will see in the decreasing distance, a light, the "light" of "Heaven" and of paradise. Once embraced by the light "the recently dead" may be greeted by the souls of dead relatives, friends, and/or radiant human or animal-like entities (Eadie, 1992; Rawling 1978; Ring 1980).

However, emerging from the tunnel and mouth of the cave, would also be a symbolic rebirth through the birth canal and womb of the earth....

11. DREAMS, ANIMAL SPIRITS AND LOST SOULS

Across time and culture, people have believed that not just humans but animals, plants and trees were sensitive, sentient, intelligent, and the abode of spirits including the souls of dead ancestors (Campbell 1988; Frazier 1950; Harris, 1993; Jung 1964; Malinowski 1948). Before hunting and killing an animal, its spirit sometimes had to be conjured forth so as to not harm it, or to ask forgiveness (Campbell 1988; Frazier 1950). The great hunters respected and paid homage to the souls and spirits of the animals they killed, and the Cro-Magnon were great hunters.

Figure: A dead hunter and a birds head. A disemboweled bison stands above him. Presumably this scene depicts the death of a hunter and the flight of his soul as symbolized by the bird.

Bird heads were commonly employed by ancient peoples including the Egyptians to depict the ascension to heaven. Eventually, bird heads were replaced by creatures with wings, e.g. angels. However, the symbolism of the birds also refers to flight, and the spirits of the dead were believed to ascend to the heavens which were filled with stars.

Figure: Goddesses with bird heads. From Libya, approximately 10,000 B.P. The bird head symbolizes the capacity for flight and thus the ability to ascend to heaven.

Be it human, animal or plant, souls were also believed capable of migrating to new abodes, and that souls could migrate from humans, to animals or plants and then back again (Campbell, 1988; Frazier, 1950; Harris, 1993; Jung, 1964; Malinowkski, 1948). The spirit left the body at death, and the body was buried in the womb of the earth, from which new life would emerge. And the liberated soul might ascend to starry vault of heaven, sometimes taking the shape and form of a bird.

Be it following death, or during a dream, sometimes the soul was believed to take on another form, such as a bird, or deer, fox, rabbit, wolf, and so on. The spirit and the soul could also hover about in human- or animal-like, ghostly vestiges, at the fringes of reality, the hinterland where day turns into night (Campbell 1988; Frazier 1950; Jung 1964; Malinowski 1954; Wilson 1951). The souls of animal's such as a wolf or eagle, could also leave the body and take on various forms including that of a woman or Man. Not just men but animals too had souls that had to be respected.

Even after death souls continued to interact with the living, and every living being possessed a soul. Hence, the ancients believed that these souls could be influenced, their behavior controlled, and, in consequence, a good hunt insured or with the assistance of a soul. And thus, deep within the womb of the Earth, the Cro-Magnon painted and paid homage to the spirits of the animal world.

Souls were also believed by ancient humans to wonder about while people sleep and dream (Brandon 1967; Frazier 1950; Harris 1993; Jung 1945, 1964; Malinowkski 1990). That is, among many different cultures and religions the soul is believed to sometimes wonder away from the body, especially while dreaming, and may engage in certain acts or interact with other souls including those of the dear but long dead and departed. These peoples believed in an afterlife and a spirit world which could be entered through a doorway of dreams. Thus, at death, the soul or spirit would be completely liberated from the body.

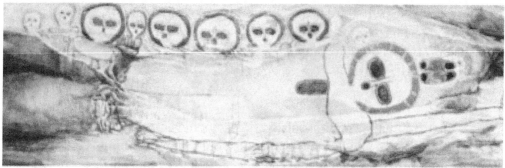

Figure: (above) Spirits and the souls (wondjinas) of the dead. (Below) Paleolithic spirits ascending to the heavens.

According to the ancients, the soul could exit the body following death and thus we see that the peoples of the Paleolithic peoples often buried their dead in sleeping positions. And, because the Cro-Magnons obviously believed in an after-life, they buried their dead with food, weapons, flowers, jewelry, clothing, pendants, rings, necklaces, multifaceted tools, head bands, beads, bracelets and so on. The Cro-Magnon were a profoundly spiritual people and they fully prepared the dead for the journey to the spirit world, equipping them so that they could live for all eternity in the land of the ancestors and the gods.

12. THE COSMOLOGY OF ANCIENT SPIRITUALITY

When humans first turned their eyes to the sun, moon, and stars to ponder the nature of existence and the cosmos, is unknown. The Cro-Magnon people were keen observers of the world around them, which they depicted with artistic majesty. The heavens were part of their world and they searched the skies for signs and observed the moon, the patterns formed by clusters of stars, and perhaps the relationship between the Earth, the sun, and the changing seasons. Although it is impossible to date cave paintings with precision, the first evidence of this awareness of the cosmic connection between Sun, Moon, Woman, Earth and the changing seasons are from the Paleolithic; symbolized in the creations of the Cro-Magnon.

12.1. GODDESS OF THE MOON Among the ancients, the Sun and the Moon were of particular importance and the Cro-Magnon observed the relationship between woman and the lunar cycle. Consider, the pregnant goddess, the Venus of Laussel, who holds the crescent moon in her hand (though others say it is a bison's horn). Although the length of a Cro-Magnon woman's menstrual cycle is unknown, it can be assumed that like modern woman she menstruated once every 28 to 29 days, which corresponds to a lunar month 29 days long, and which averages out to 13 menstrual cycles in a solar year. And not just menstruation, but pregnancy is linked to the phases of the moon.

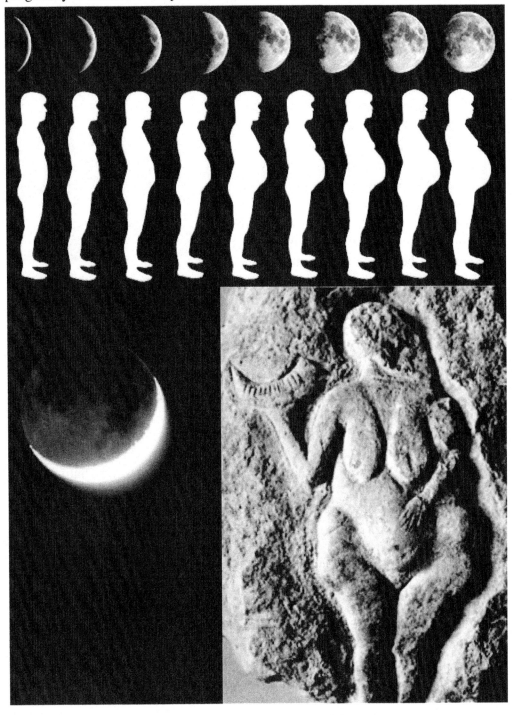

Figure: The entrance to the underground Upper Paleolithic cathedral. The Chauvet cave. Note sign of the cross. From Chauvet et al., (1996). Dawn of Art: The Chauvet Cave. Henry H. Adams. NY

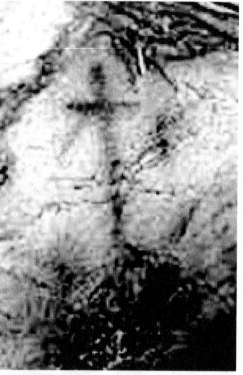

12.2. THE FOUR CORNERS OF THE SOLAR CLOCK.

When the Cro-Magnon turned their eyes to the heavens, seeking to peer beyond the mystery that separated this world from the next, they observed the sun. With a brain one third larger than modern humans, and given their tremendous power of observation, it can be predicted these ancient people would have associated the movement of the sun with the changing seasons which effected the behavior of animals, the growth of plants, and the climate and weather; all of which are directly associated with cyclic alterations in the position of the sun and the length of a single day over the course of a solar year which is equal to 13 moons.

The four seasons, marked by two solstices and the two equinoxes have been symbolized by most ancient cultures with the sign of the cross, e.g. the "four corners" of the world and the heavens. The "sign of the cross" generally signifies religious or cosmic significance. The Cro-Magnon also venerated the sign of the cross, the first evidence of which, an engraved cross, is at least 60,000 years old (Mellars, 1989). Yet another cross, was painted in bold red ochre upon the entryway to the Chauvet Cave, dated to over 30,000 years ago (Chauvet et al., 1996).

The illusion of movement of the Sun, from north to south, and then back again, in synchrony with the waxing and waning of the four seasons, is due to the changing tilt and inclination of the Earth's axis, as it spins and orbits the sun. Thus over a span of 13 moons, it appears to an observer that the days become shorter and then longer and then shorter again as the sun moves from north to south, crosses the equator, and then stops, and heads back north again, only to stop, and then to again head south, crossing the equator only to again stop and head north again. The two crossings each year, over the equator (in March and September) are referred to as equinoxes and refers to the days and nights being of equal length. The two time periods in which the sun appears to stop its movement, before reversing course (June and December), are referred to as solstices—the "sun standing still."

The sun was recognized by ancient astronomer priests, as a source of light and life-giving heat, and as a keeper of time, like the hands ticking across the face of a cosmic clock. Because of the scientific, religious, and cosmological significance of the sun, ancient peoples, in consequence, often erected and oriented their religious temples to face and point either to the rising sun on the day of the solstice (that is, in a southwest—northeast axis), or to face the rising sun on the day of the equinox (an east-west axis). For example, the ancient temples and pyramids in Egypt were oriented to the solstices, whereas the Temple of Solomon faced the rising sun on the day of the equinox.

Thus the sign of the cross is linked to the heavens and to the sun. Understanding the heavens and the sun, has been a common astronomical method of divining the will of the gods, and for navigation, localization, and calculation: these celestial symbols have heavenly significance.

Regardless of time and culture, from the Aztecs, Mayas, American Indians, Romans, Greeks, Africans, Christians, Cro-Magnons, Egyptians (the key of life), and so on, the cross consistently appears in a mystical context, and/or is attributed tremendous cosmic significance (Budge,1994; Campbell, 1988; Joseph, 2000a; Jung, 1964). The sign of the cross was the ideogram of the goddess "An", the Sumerian giver of all life from which rained down the seeds of life on all worlds including the worlds of the gods. An of the cross gave life to the gods, and to woman and man.

Figure: God Seb supporting Goddess Nut who represents heaven. Key of life: ring with a cross.

The symbol of the cross is in fact associated with innumerable gods and goddesses, including Anu of the ancient Egyptians, the Egyptian God Seb, the Goddess Nut, the God Horus (the hawk), as well as Christ and the Mayan and Aztec God, Quetzocoatl. For example, like the Catholics, the Mayas and Aztecs adorned their temples with the sign of the cross. Quetzocoatl, like Jesus, was a god of the cross.

Figure: Quetzocoatl god of the cross. Round shield encircling the cross represents the sun.

In China the equilateral cross is represented as within a square which represents the Earth, the meaning of which is: "God made the Earth in the form of a cross." It is noteworthy that the Chinese cross-in-a-box can also be likened to the swastika—also referred to as the "gammadion" which is one of the names of the Lord God: "Tetragammadion." The cross, in fact forms a series of boxes when aligned from top to bottom or side by side, and cross-hatchings such as these were carved on stone over 60,000 years ago.

Figure: Ochre etched with crosses, forming a series of cross-hatchings, dating to 77,000 years ago.

Quantum Physics, Neuroscience of Mind

Among the ancient, the sign of the cross, represented the journey of the sun across the four corners of the heavens. The Cro-Magnon adorned the entrance and the walls of their underground cathedrals with the sign of the cross, which indicates this symbol was of profound cosmic significance. However, that some of the Cro-Magnon depictions of animal-headed men have also been found facing the cross, may also pertain to the heavens: the patterns formed by stars, which today are refereed to as "constellations."

12.3. THE CONSTELLATION OF VIRGO There is nothing "virginal" about the constellation of Virgo. The pattern can be likened to a woman in lying on her back with an arm behind her head, and this may have been the visage which stirred the imagination of the Cro-Magnon.

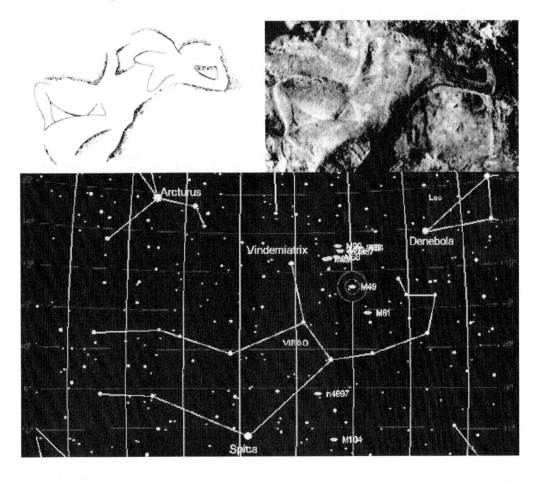

Figure: Cro-Magnon goddess, depicting the constellation of Virgo. La Magdelain cave.

12.4. THE PLEIADES AND THE CONSTELLATIONS OF TAURUS AND OSIRIS

It would be unreasonable to assume that the Cro-Magnon would not have observed the heavens or the illusory patterns formed by the alignment of various stars. Depictions of the various constellations, such as Taurus and Orion, and "mythologies" surrounding them, are of great antiquity, and it appears that similar patterns were observed by the Cro-Magnon people.

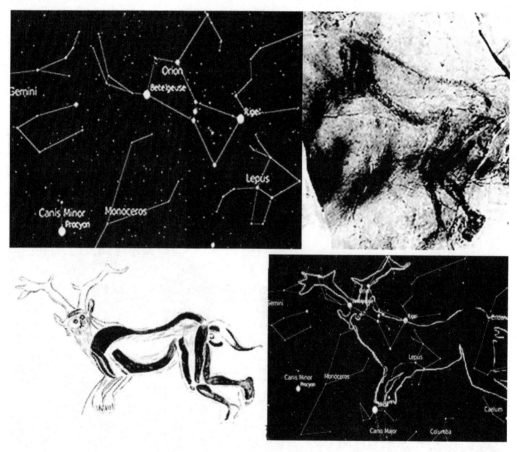

Figures: (Upper Right / Lower Left) The "Sorcerer" Trois-Frères cave. (Upper Left / Lower Right) Constellation of Orion/Osiris.

Consider, for example the "Sorcerers" or "Shamans" wearing the horns of a bull, and possibly representing the constellation of Taurus; a symbol which appears repeatedly in Lascaux, the "Hall of the Bulls" and in the deep recesses of other underground cathedrals dated from 18,000 to 30,000 B.P. And above the back of one of these charging bulls, appears a grouping of dots, or stars, which many authors believe may represent the Pleiades which is associated with Taurus. These Paleolithic paintings of the bull appear to be the earliest representation of the Taurus constellation.

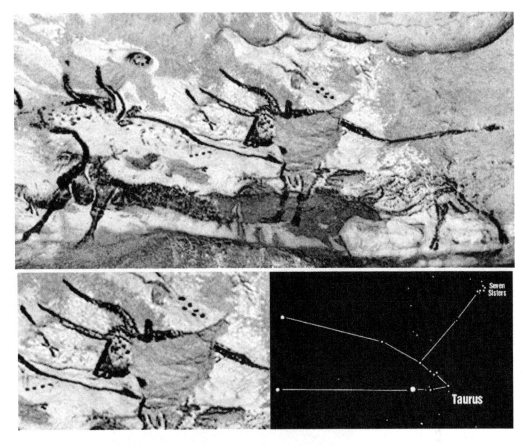

Figure: (Top) The main freeze of the bulls in the Lascaux Cave in Dordogne. There is a group of dots on the back of the great bull (Taurus) which may represent six of the seven stars of the Pleiades (the seven sisters). As stars are also in motion, not all would be aligned or as bright or dim today, as was the case 20,000 to 30,000 years ago.

In the "modern" sky, the constellation of Osiris/Orion the hunter, faces Taurus, the bull; and these starry patterns would not have been profoundly different 20,000 to 30,000 years ago. In ancient Egypt, dating back to the earliest dynasties (Griffiths 1980), Osiris was the god of death and of fertility and rebirth, who wore a distinctive crown with two horns (later symbolized as ostrich feathers at either side). He was the brother and husband of Isis. According to myth, Osiris was killed by Set (the destroyer) and dismembered. Isis recovered all of his body, except his penis. After his death she becomes pregnant by Osiris. The Kings of Egypt were believed to ascend to heaven to join with Osiris in death and thereby inherit eternal life and rebirth, symbolized by the star Sirius (Redford 2003). The Egyptian "King list" (The Turin King List) goes backward in time, 30,000 years ago to an age referred to as the "dynasty of gods" which was followed by a "dynasty of demi-gods" and then dynasties of humans (Smith 1872/2005).

Over 20,000 years ago, the 6ft tall Cro-Magnon, with their massive brain one third larger than modern humans, painted a hunter with two horns who had been killed. And just as the constellation of Orion the hunter faces Taurus, so too does the dead Cro-Magnon hunter who has dismembered/disemboweled the raging bull. And below and beneath the dead Cro-Magnon hunter, another bird, symbol of rebirth, and perhaps symbolizing the star Sirius.

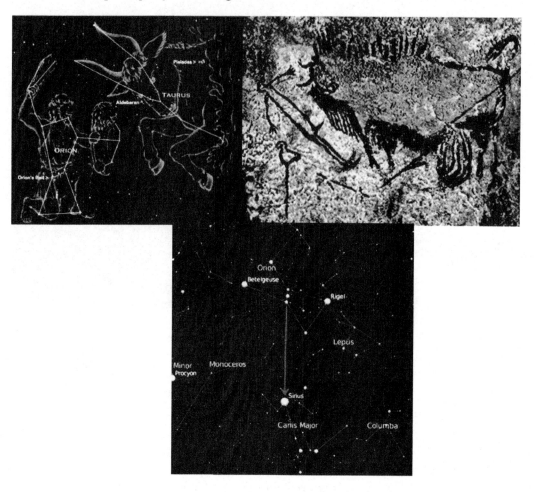

Figures: The constellation of Osiris (Orion the hunter) in Egyptian mythology is the god of the dead who was dismembered; but also represents resurrection and eternal life as signified by the star Sirius. (Upper Right) Constellation of Osiris/Orion and Taurus. (Upper Left) Cave painting. Lascaux. The dead (bird-headed or two horned) hunter killed by a bull whom he disemboweled. (Bottom) Constellation of Orion/Osiris in relation to Sirius.

13. CONCLUSIONS

Complex mortuary rituals and belief in the transmigration of the soul, of a world beyond the grave, has been a human characteristic for at least 100,000 years. The emergence of spiritual consciousness and its symbolism, is directly linked to the evolution of the temporal and frontal lobes and to the Neanderthal and Cro-Magnon.

These ancient peoples were capable of experiencing love, fear, and mystical awe, and they carefully buried those they loved and lost. They believed in spirits and ghosts which dwelled in a heavenly land of dreams, and interned their dead in sleeping positions and with tools, ornaments and flowers. By 30,000 years ago, and with the expansion of the frontal lobes, they created symbolic rituals to help them understand and gain control over the spiritual realms, and created signs and symbols which could generate feelings of awe regardless of time or culture.

Because they believed souls ascended to the heavens, the people of the Paleolithic searched the heavens for signs. They observed and symbolically depicted the association between woman and the moon, patterns formed by stars, and the relationship between Earth, the sun, and the four seasons.

The ancestry and origins of the Cro-Magnon peoples, are completely unknown. There are no transitional forms that link them with Neanderthals or the still primitive "early modern" peoples of the Middle Paleolithic who were decidedly more archaic in appearance as compared to Cro-Magnons. Neanderthals did not evolve into Cro-Magnons, and they coexisted for almost 15,000 years, until finally the Neanderthals disappeared from the face of the Earth, around 30,000 years B.P. (Mellars, 1996). Indeed, the Neanderthals were of a completely different race; and not just physically, but genetically, for when they died out, so too did their genetic heritage and almost all traces of their DNA (Conrad & Richter 2011; Harvati & Harrison 2010).

Since Cro-Magnons shared the planet with Neanderthals during overlapping time periods it certainly seems reasonable to assume that the technologically and intellectually superior Cro-Magnons probably engaged in wide spread ethnic cleansing and exterminated the rather short (5ft 4in.), sloped-headed, heavily muscled Neanderthals, eradicating all but hybrids from the face of the Earth, some 35,000 to 28,000 years ago.

Presumably, the evolutionary lineage of the Cro-Magnon is linked to the evolution of "early modern" archaic humans who first appeared in what is now the Middle East around 100,000 years ago. In contrast, to Neanderthals, the frontal portions of

Cosmology of Consciousness

the craniums of "early modern" archaics, were rounded, indicating an expansion of the frontal lobes (Joseph 1996). In addition, these archaics began engaging in mortuary rites before the Neanderthals. For example, archaic H. sapiens and "early moderns" were carefully buried in Qafzeh, near Nazareth and in the Mt. Carmel, Mugharetes-Skhul caves on the Western coast of the Middle East over 90,000 to 98,000 years ago (McCown 1937; Smirnov 1989; Trinkaus 1986). This includes a Qafzeh mother and child who were buried together, and an infant who was buried holding the antlers of a fallow deer across his chest. In a nearby site equally as old (i.e. Skhul), yet another was buried with the mandible of a boar held in his hands, whereas an adult had stone tools placed by his side (Belfer-Cohen and Hovers 1992; McCown 1937). "Early modern," and other "archaic" Homo sapiens commonly buried infants, children, and adults with tools, grave offerings, and animal bones.

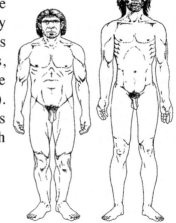

Figure: (Left) Neanderthal. (Right) Cro-Magnon.

However, it was not until the evolution of the Cro-Magnon and the expansion of the frontal lobes that symbolic representations of religious and spiritual feelings literally became an art. The spiritual belief systems of the Cro-Magnon and other peoples of the Upper Paleolithic, completely outstripped those of their predecessors in complexity, originality, and artistic and symbolic expression. Hence, the Cro-Magnon conception of, and ability to symbolically express the spirit world, became much more complex as well, undergoing what has been described as a "symbolic explosion". As the brain and man and woman evolved, so too did their spiritual beliefs.

With a massive frontal lobe and a brain one third larger than modern humans, the 6ft tall Cro-Magnons were intellectual giants, as the remnants of their creations attest. What they might have been capable of mentally, what they may have achieved is unknown to us, except indirectly, through what today is classified as "myth."

What is known for fact is the people of the Paleolithic were among the greatest hunters, craftsmen, and artists to have walked this Earth. It is these people who were the first to develop complex beliefs involving spirits, souls, sorcerers, shamans, goddesses, and the moon, and sun, and the stars which shine in the darkness of night.

Spiritual consciousness first began to evolve 100,000 years ago. It is this consciousness of the spirit, and belief in the transcendence of the soul, which gave birth to the first heavenly cosmologies over 20,000 years ago.

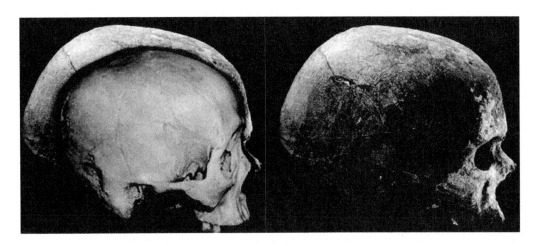

Figure: Comparison of modern human skull superimposed on Cro-Magnon skull (Left). Cro-Magnon skull (Right). The Cro-Magnon brain was 1/3 larger on average, than the modern human brain.

References

Akazawa , T & Muhesen, S. (2002). Neanderthal Burials. KW Publications Ltd.

Amaral, D. G., Price, J. L., Pitkanen, A., & Thomas, S. (1992). Anatomical organization of the primate amygdaloid complex. In J. P. Aggleton (Ed.). The Amygdala. (Wiley. New York.

Bandi, H. G. (1961). Art of the Stone Age. New York, Crown PUblishers, New York.

Bear, D. M. (1979). Temporal lobe epilepsy: A sydnrome of sensory-limbic hyperconnexion. Cortex, 15, 357-384.

Belfer-Cohen, A., & E.Hovers, (1992). In the eye of the beholder: Mousterian and Natufian burials in the levant. Current Anthropology 33: 463-471.

Breuil. H. (1952). Four hundred centuries of cave art. Montignac.

Budge, W. (1994). The Book of the Dead. New Jersey, Carol.

Butzer, K. (1982). Geomorphology and sediment stratiagraphy, in The Middle Stone Age at Klasies River Mouth in South Africa. Edited by R. Singer and J. Wymer. Chicago: University of Chicago Press.

Binford, L. (1981). Bones: Ancient Men & Modern Myths. Academic Press, NY

Binford, S. R. (1973). Interassemblage variability--the Mousterian and the 'functional' argument. In The explanation of culture change. Models in prehistory. edited by C. Renfrew. Pittsburgh: Pittsburgh U. Press.

Binford S. R. (1982). Rethinking the Middle/Upper Paleolithic transition. Current Anthropology 23: 177-181.

Blinkov, S. M., & Glezer, I. I. (1968). The human brain in figures and tables. New York: Plenum.

Campbell, J. (1988) Historical Atlas of World Mythology. New York, Harper & Row.

Cartwright , R. (2010) The Twenty-four Hour Mind: The Role of Sleep and Dreaming in Our Emotional Lives. Oxford University Press.

Chauvet, J-M., Deschamps, E. B. & Hillaire, C. (1996) Dawn of Art: The Chauvet Cave. H.N. Abrams.

Clark, G. (1967) The stone age hunters. Thames & Hudson.

Clark, J. D., & Harris, J. W. K. (1985). Fire and its role in early hominid lifeways. African Archaeology Review, 3, 3-27.

Conrad, N. J., & Richter, J. (2011). Neanderthal Lifeways, Subsistence and Technology. Springer.

Daly, D. (1958) Ictal affect. American Journal of Psychiatry, 115, 97-108.

Day, M. H. (1996). Guide to Fossil Man. University of Chicago Press, Chicago.

Dennell, R. (1985). European prehistory. London, Academic Press.

Eadie, B. J. (1992). Embraced by the light. California, Gold Leaf Press.

Frazier, J. G. (1950). The golden bough. Macmillan, New York.

Freud, S. (1900). The interpretation of dreams. Standard Edition (Vol 5). London: Hogarth Press.

Gloor, P. (1997). The Temporal Lobes and Limbic System. Oxford University Press. New York.

Gowlett, J. (1984). Ascent to civlization. New York: Knopf.

Gowlett, J.A. (1981). Early archaeological sites, hominid remains and traces of fire from Chesowanja, Kenya. Nature, 294, 125-129.

Griffiths, J. G. (1980). The Origins of Osiris and His Cult. Brill.

Halgren, E. (1992). Emotional neurophysiology of the amygdala within the context of human cognition. In J. P. Aggleton (Ed.). The Amygdala. New York, Wiley-Liss.

Halgren, E., Walter, R. D., Cherlow, D. G., & Crandal, P. H. (1978). Mental phenomenoa evoked by electrical stimualtion of the human hippocampal formation and amygdala, Brain, 101, 83-117.

Cosmology of Consciousness

Harold, F. B. (1980). A comparative analysis of Eurasian Palaeolithic burials. World Archaeology 12: 195-211.

Harold, F. B. (1989). Mousterian, Chatelperronian, and Early Aurignacian in Western Europe: Continuity or disconuity?" In P. Mellars & C. B. Stringer (eds). The human revolution: Behavioral and biological perspectives on the origins of modern humans, vol 1.. Edinburgh: Edinburgh University Press.

Harris, M. (1993) Why we became religious and the evolution of the spirit world. In Lehmann, A. C. & Myers, J. E. (Eds) Magic, Witchcraft, and Religion. Mountain View: Mayfield.

Hasselmo, M. E., Rolls, E. T., & Baylis, G. C. (1989). The role ofexpression and identity in the face-selective responses of neurons in thetemporal visual cortex of the monkey. Behavioral Brain Research, 32,203-218.

Harvati, K., & Harrison, T. (2010). Neanderthals Revisited. Springer.

Hayden, B. (1993). The cultural capacities of Neandertals: A review and re-evaluation. Journal of Human Evolution 24: 113-146.

Holloway, R. L. (1988) Brain. In: Tattersall, I., Delson, E., Van Couvering, J. (Eds.) Encyclopedia of human evolution and prehistory. New York: Garland.

Joseph, R. (1982). The Neuropsychology of Development. Hemispheric Laterality, Limbic Language, the Origin of Thought. Journal of Clinical Psychology, 44 4-33.

Joseph, R. (1986). Confabulation and delusional denial: Frontal lobe and lateralized influences. Journal of Clinical Psychology, 42, 845-860.

Joseph, R. (1988) The Right Cerebral Hemisphere: Emotion, Music, Visual-Spatial Skills, Body Image, Dreams, and Awareness. Journal of Clinical Psychology, 44, 630-673.

Joseph, R. (1990a) The temporal lobes. In A. E. Puente and C. R. Reynolds (series editors). Critical Issues in Neuropsychology. Neuropsychology, Neuropsychiatry, Behavioral Neurology. Plenum, New York.

Joseph, R. (1990b) The frontal lobes. In A. E. Puente and C. R. Reynolds (series editors). Critical Issues in Neuropsychology. Neuropsychology, Neuropsychiatry, Behavioral Neurology. Plenum, New York.

Joseph, R. (1992) The Limbic System: Emotion, Laterality, and Unconscious Mind. The Psychoanalytic Review, 79, 405-456.

Joseph, R. (1993) The Naked Neuron. Evolution and the languages of the body and the brain. Plenum. New York.

Joseph, R. (1994) The limbic system and the foundations of emotional experience. In V. S. Ramachandran (Ed). Encyclopedia of Human Behavior. San Diego, Academic Press.

Joseph, R. (1996). Neuropsychiatry, Neuropsychology, Clinical Neuroscience, 2nd Edition. 21 chapters, 864 pages. Williams & Wilkins, Baltimore.

Joseph, R. (1998a). The limbic system. In H.S. Friedman (ed.), Encyclopedia of Human health, Academic Press. San Diego.

Joseph, R. (1998b). Traumatic amnesia, repression, and hippocampal injury due to corticosteroid and enkephalin secretion. Child Psychiatry and Human Development. 29, 169-186.

Joseph, R. (1999a). Environmental influences on neural plasticity, the limbic system, and emotional development and attachment, Child Psychiatry and Human Development. 29, 187-203.

Joseph, R. (1999b). The neurology of traumatic "dissociative" amnesia. Commentary and literature review. Child Abuse & Neglect. 23, 715-727.

Joseph, R. (1999c). Frontal lobe psychopathology: Mania, depression, aphasia, confabulation, catatonia, perseveration, obsessive compulsions, schizophrenia. journal of Psychiatry, 62, 138-172.

Joseph, R. (2000a). The Transmitter to God. University Press California.

Joseph, R. (2000b). The evolution of sex differences in language, sexuality, and visual spatial skills. Archives of Sexual Behavior, 29, 35-66.

Joseph, R. (2000c). Fetal brain behavioral cognitive development. Developmental Review, 20, 81-98.

Joseph, R. (2001). The Limbic System and the Soul: Evolution and the Neuroanatomy of Religious Experience. Zygon, the Journal of Religion & Science, 36, 105-136.

Joseph, R. (2002). NeuroTheology: Brain, Science, Spirituality, Religious Experience. University Press.

Joseph, R. (2011a). Dreams and Hallucinations: Lifting the Veil to Multiple Perceptual Realities, Cosmology, 14, In press.

Joseph, R. (2011). The neuroanatomy of free will: Loss of will, against the will "alien hand", Journal of Cosmology, 14, In press.

Jung, C. G. (1945). On the nature of dreams. (Translated by R.F.C. Hull.), The collected works of C. G. Jung, (pp.473-507). Princeton: Princeton University Press.

Jung, C. G. (1964). Man and his symbols. New York: Dell.

Kawashima, R., Sugiura, M., Kato, T., et al., (1999). The human amygdala plays an important role in gaze monitoring. Brain, 122, 779-783.

Kling. A. S. & Brothers, L. A. (1992). The amygdala and social behavior. In J. P. Aggleton (Ed.). The Amygdala. New York, Wiley-Liss.

Kurten, B. (1976). The cave bear story. New York: Columbia University Press.

Leroi-Gourhan, A. (1964.) Treasure of prehistoric art. New York: H. N. Abrams.

Leroi-Gourhan, A. (1982). The archaeology of Lascauz Cave. Scientific American 24: 104-112.

MacLean, P. (1990). The Evolution of the Triune Brain. New York, Plenum.

Malinowski, B. (1954) Magic, Science and Religion. New York. Doubleday.

McCown, T. (1937). Mugharet es-Skhul: Description and excavation, in The stone age of Mount Carmel. Edited by D. A. E. Garrod and D. Bate. Oxford: Clarendon Press.

Mellars, P. (1989). Major issues in the emergence of modern humans. Current Anthropology 30: 349-385.

Mellars, P. (1996) The Neanderthal legacy. Princeton University Press.

Mellars, P. (1998). The fate of the Neanderthals. Nature 395, 539-540.

Mesulam, M. M. (1981) Dissociative states with abnormal temporal lobe EEG: Multiple personality and the illusion of possession. Archives of General Psychiatry, 38, 176-181.

Moody, R. (1977). Life after life. Georgia, Mockingbird Books.

Morris, J. S., Frith, C. D., Perett, D. I., Rowland, D., Young, A. W., Calder, A. J., & Colan, R. J. (1996). A differential neural response in the human amygdala to fearful and happy facial expression. Nature, 383, 812-815.

Mullan, S., & Penfield, W. (1959). Epilepsy and visual halluciantions. Archives of Neurology and Psychiatry, 81, 269-281.

Neihardt, J. G. & Black Elk, (1979). Black Elk speaks. Lincoln. U. Nebraska Press.

emchin, A. A., Whitehouse, M.J., Menneken, M., Geisler, T., Pidgeon, R.T., Wilde, S. A. (2008). A light carbon reservoir recorded in zircon-hosted diamond from the Jack Hills. Nature 454, 92-95.

Noyes, R., & Kletti, R. (1977) Depersonalization in response to life threatening danger. Comprehensive Psychiatry, 18, 375-384.

Parson, E. R. (1988). Post-traumatic self disorders (PTsfD): Theoretical and practical considerations in psychotherapy of Vietnam War Veterans. In J. P. Wilson, Z. Harel, & B. Kahana (Eds). Human Adaptation to Extreme Stress. New York, Plenum.

Pena-Casanova, J., & Roig-Rovira, T. (1985). Optic aphasia, optic apraxia, and loss of dreaming. Brain and Language, 26, 63-71.

Penfield, W., & Perot, P. (1963). The brains record of auditory and visual experience. Brain, 86, 595-695.

Perryman, K. M., Kling, A. s., & Lloyd, R. L. (1987). Differential effects of inferior temporal cortex lesions upon visual and auditory-evoked potentials in the amygdala of the squirrel monkey. Behavioral and Neural Biology, 47, 73-79.

Petrides, M., & Pandya, D. N. (1999). Dorsolateral prefrontal cortex: comparative cytoarchitectonic analysis in the human and the macaque brain and corticocortical connection patterns. European Journal of Neuroscience 11.1011–1036.

Petrides, M., & Pandya, D. N. (2001). Comparative cytoarchitectonic analysis of the human and the macaque ventrolateral prefrontal cortex and corticocortical connection patterns in the monkey. European Journal of Neuroscience 16.291–310.

Prideaux, T. (1973). Cro-Magnon. New York: Time-Life.

Rawlings, M. (1978). Beyond deaths door. London, Sheldon Press.

Redford, D. B. (2003). The Oxford Guide: Essential Guide to Egyptian Mythology, Berkley.

Rightmire, G. P. (1984). Homo sapiens in Sub-Saharan Africa, In F. H. Smith and F. Spencer (eds). The origins of modern humans: A world survey of the fossil evidence. New York: Alan R. Liss.

Ring, K. (1980). Life at death. New York, Coward, McCann & Geoghegan.

Roginskii Y. Y., & Lewin S. S. (1955). Fundamentals of Anthropology. Moscow: Moscow University Press.

Rolls, E. T. (1984). Neurons in the cortex of the temporal lobe and in the amygdala of the monkey with responses selective for faces. Human Neurobiology, 3, 209-222.

Rolls, E. T. (1992). Neurophysiology and functions of the primate amygdala. In J. P. Aggleton (Ed.). The Amygdala. New York, Wiley-Liss.

Sabom, M. B. (1982). Recollections on death. New York, Harper & Row.

Sawa, M., & Delgado, J. M. R. (1963). Amygdala unitary activity in the unrestrained cat. Electroencephalography and Clinical Neurophysiology, 15, 637-650.

Schenk, L., & Bear, D. (1981) Multiple personality and related dissociative phenomenon in patients with temporal lobe epilepsy. American Journal of Psychiatry, 138, 1311-1316.

Schutze, I., Knuepfer, M. M., Eismann, A., Stumpf, H., & Stock, G. (1987). Sensory input to single neurons in the amygdala of the cat. Experimental Neurology, 97, 499-515.

Slater, E. & Beard, A.W. (1963). The schizophrenia-like psychoses of epilepsy. British Journal of Psychiatry, 109, 95-112.

Schwarcz, A. et al. (1988). ESR dates for the hominid burial site of Qafzeh. Journal of Human Evolution 17: 733-737.

Smirnov, Y. A. (1989). On the evidence for Neandertal burial. Current Anthropology 30: 324.

Smith, G. A. (1872/2005). Chaldean Account of Genesis (Whittingham & Wilkins, London, 1872). Adamant Media Corporation (2005).

Solecki, R. (1971). Shanidar: The first flower people. New York: Knopf.

Subirana, A., & Oller-Daurelia, L. (1953). The seizures with a feeling of paradisiacal happiness as the onset of certain temporal symptomatic epilepsies. Congres Neurologique International. Lisbonne, 4, 246-250.

Tarachow, S. (1941). The clinical value of hallucinations in localizing brain tumors. American Journal of Psychiatry, 99, 1434-1442.

Taylor, D. C. (1972). Mental state and temporal lobe epilepsy. Epilepsia, 13, 727-765.

Tilney, F. (1928). The brain from ape to man. New York: P. B. Hoeber.

Tobias, P. V. (1971). The Brain in Hominid Evolution. Columbia University Press, New York.

Trimble, M. R. (1991). The psychoses of epilepsy. New York, Raven Press.

Trinkaus, E. (1986). The Neanderthals and modern human origins. Annual Review of Anthropology 15: 193-211.

Turner, B. H. Mishkin, M. & Knapp, M. (1980). Organization of the amygdalopetal projections from modality-specific cortical association areas in the monkey. Journal of Comparative Neurology, 191, 515-543.

Ursin H., & Kaada, B. R. (1960). Functional localization within the amygdaloid complex in the cat. Electroencephalography and Clinical Neurophysiology, 12, 1-20.

Weil, A. (1929). Measurements of cerebral and cerebellar surfaces. American Journal of Physical Anthropology 13: 69-90.

Weingarten, S. M., Cherlow, D. G. & Holmgren. E. (1977). The relationship of hallucinations to depth structures of the temporal lobe. Acta Neurochirugica 24: 199-216.

Williams, D. (1956). The structure of emotions reflected in epileptic experiences. Brain, 79, 29-67.

Wilson, I. (1987). The after death experience. New York, Morrow.

Wilson, J. A. (1951) The culture of ancient Egypt. Chicago, U. Chicago Press.

Wolpoff, M. H. (1980), Paleo-Anthropology. New York, Knopf.

Alien Life and Quantum Consciousness

Randy D. Allen, Ph.D.,
Department of Biochemistry and Molecular Biology, Oklahoma State University.

Dr. Steven Hawking (2010) argues that alien life almost surely exists and warns that human encounters with aliens could well end badly for us. In the absence of actual evidence for the presence of extraterrestrial life, much less for its intentions, I would say that Dr. Hawking's guesses are as good as anyone else's. I guess that it is equally likely that extraterrestrial life does not exist either because it never arose or, if it did arise, was soon eliminated. Another possibility is that some sort of primitive photosynthetic or chemosynthetic organisms exist elsewhere in the Universe. If, as on Earth, the primary form of energy available on other planets is radiation from a nearby star, it seems likely that, as on Earth, photosynthetic organisms must predominate. It follows that consumers of these producers may well have also evolved. However, life that exists away from Earth will not necessarily use the same chemistry. If, as we generally assume, water is required for life, then the range of possible chemistries is constrained but there is absolutely no reason to assume that anything remotely close to plants or animals, much less humans, exists elsewhere. Although dissimilarities in chemistry could limit our ability to analyze or even detect extraterrestrials, it could also protect us from harm. An alien organism without proteins, for example, may have a difficult time digesting us. If, on the other hand, we were to encounter alien life forms that ravenously consume all of the energy containing chemical compounds that they can get their "hands" on, it would indeed be tragic for us, but a good meal for them.

Life as we know it is based on chemistry but, what if life elsewhere is based, not on chemistry but on quantum mechanics? Imagine alien life forms that can manipulate subatomic particles like our cells manipulate chemical compounds. Humans have existed as a species for less than a million years and we are, as far as we know, the only species on Earth that has even the vaguest notion of physics. We only discovered the atom and learned to unleash its power within the last century. Our understanding of quantum mechanics is rudimentary, at best, yet we are on the verge of developing practical quantum computers that promise virtually unlimited computational power. It is conceivable that, in the billions of years since the Big Bang, other organisms evolved at some time and some place that have already mastered quantum mechanics. Let's say that intelligent, social, organisms with chemically-based metabolism, fundamentally not unlike ourselves, evolved on a planet somewhere in the universe. Their unquenchable curiosity about the universe (or, like us, their unquenchable desire to exploit it) led them to develop efficient quantum computers. They realized that, with such computers, the whole of their existence could be computerized, all memories and life experiences, all emotions and motivations, could be transferred to a collective "quantum brain". In effect, their "species", though biologically extinct, could become immortal. No more inefficient metabolism requiring huge energy input, no chemically derived bodies to wear out, no reproduction, no death, no taxes. Just supermassively parallel collective consciousness with unlimited capabilities. Perhaps, through super symmetry or entanglement, they can "see" or "feel" the entire universe. Maybe, they've gained the ability to manipulate elementary particles and can control its evolution and its fate. They would have become, by any human definition, Gods.

It's conceivable that quantum capabilities evolved multiple times throughout the universe, each new member of the quantum club bringing a novel dimension of consciousness, along with a few billion additional "neighbors" to get to know. With no need to compete for resources, quantum beings are probably peaceful and only want the best for the Universe and its inhabitants. Maybe they are aware of our existence but don't care about us, much as we ignore most of the "lower" organisms that surround us. Alternatively, perhaps they have noted our biological, social and technological evolution and realize that we humans may well join their ranks someday and become quantum beings ourselves.

The possibility of evolving a quantum consciousness of course, depends on numerous variables, and requires that we are not first exterminated by an asteroid impact, a nearby supernova, or gigantic volcanic eruptions, or our civilization is not decimated by global warfare over resource scarcity exacerbated by climate change. Then there is the possibility we might simply lose our scientific impetus through loss of political support for basic research and let our chance for immortality slip away.

Reference

Hawking, S. (2010). Quoted in: Stephen Hawking: alien life is out there, scientist warns. Telegraph, April 25; http://www.telegraph.co.uk/science/space/7631252/ Stephen-Hawkingalien- life-is-out-there-scientist-warns.html

Evolution of Consciousness in the Ancient Corners of the Cosmos

Rhawn Joseph, Ph.D.
Emeritus, Brain Research Laboratory, California

Abstract

Data from genetics, microbiology, astrobiology, astrophysics and cosmology, leads invariably to the conclusion that life has evolved on innumerable planets, including on worlds much older than our own. This article speculates on the nature and evolution of these life forms, and how their intelligence, brains and minds may have evolved. This theoretical article addresses a variety of issues such as: What might be the nature of life on planets with a chemistry and environment completely unlike our own? If complex life exists on worlds billions of years older than Earth, how and in what way might it have evolved? Consider our own planet. If we don't die out and continue to evolve, and if science marches on, what might humans be like a million years from now? What about ten million, a hundred million, or a billion years from now? Might they have 10 layers of neocortex instead of six? Might the lobes of the brain have greatly expanded conferring intellectual and perceptual abilities which completely dwarf our own? Might they genetically engineer their own evolution? There are stars which shine in the darkness of night which were born billions of years before the Earth was formed. The nature and evolution of life on these ancient worlds is presented.

1. We Are Not Alone

Do Extra-terrestrials exist? There are five lines of evidence which indicate the answer is "yes."

1. In 1970 lunar soil samples were returned to Earth by the Luna 16 spacecraft in a hermetically sealed container and photographed (Rode et al., 1979). The photographs were later examined by Drs. Stanislav Zhmur, and Lyudmila M. Gerasimenko, who identified what they believed to be microfossils of coccoidal bacteria which resembled Siderococcus or Sulfolobus (Klyce, 2000; Zhmur

and Gerasimenko, 1999). A third fossilized impression from the lunar surface resembles a spiral filamentous micro-Ediacaran (Joseph & Schild 2010), a species which became extinct over 500,000 years ago. In 2009, this author shared this photograph with five world-renowned experts in Cambrian and Pre-Cambrian fauna, and four of the 5 identified it as a microfossil, but too small to be an Ediacaran.

2. In 1971, a TV camera from the lunar Surveyor Space Craft was retrieved by Apollo 12 astronauts, after sitting 3 years on the moon, and a single bacterium (Streptococcus mitis) was found within (Mitchell & Ellis, 1971). In addition, the lunar camera was discovered to be covered with a film of "organic material of unknown origin" (Flory and Simoneit, 1972; Simoneit and Burlingame, 1971). The possibility of contamination prior to sending the camera to the moon, or after it was returned, was ruled out by the scientists who made this discovery. It is impossible that the microbe was the result of some other form of contamination, such as a sneeze or cough. Since a droplet of saliva contains an average of 750 million organisms, if contamination of the lunar TV camera was due to a scientist's inadvertent cough or sneeze, a multitude of related bacteria, and a "representation of the entire microbial population would be expected," rather than a single species and a single organism (Mitchell & Ellis, 1971). Moreover, this Streptococcus mitis was dormant, but came back to life.

3. There is evidence of extant life on Mars as detected by the 1976 Viking Mission Labeled Release experiment, which exploited the sensitivity of 14C respirometry and obtained positive responses at Viking 1 and 2 sites on Mars. The results indicated the possibility of living microorganisms on the red planet (Levin 2010;Levin & Straat 1976).

4. Microfossils resembling various species of bacteria, including cyanobacteria have been repeatedly discovered in meteors older than this solar system (Claus and Nagy 1961; Hoover 1998, 2006, 2011; Pflug 1984; Nagy et al. 1961,1963a,b; Zhmur and Gerasimenko 1999; Zhmur et al. 1997)

5. Two separate teams of scientists have determined, based on a genomic analysis, that DNA-based life has a genetic ancestry leading backwards in time over 10 billion years (Joseph & Wickramasinghe 2011; Sharov 2009), which is twice the age of Earth. Specifically, Joseph and Wickramasinghe (2011) determined, based on genetic analysis and estimates as to the length of time that must elapse between whole genome duplications and the divergence of species, that the first gene was fashioned between 10.4<14.5 bya.

2. Life is Everywhere?

None of these discoveries briefly mentioned above provide direct evidence for complex extra-terrestrial eukaryotic life. Nevertheless, the recent discoveries reported by Joseph, Wickramasinghe (2011) and Hoover (2011), coupled with a wealth of data from genetics, microbiology, and astrobiology detailed in the edited text, "The Biological Big Bang," (Wickramasinghe 2011), leads inescapably to the conclusion we are not alone and that complex life could have evolved on innumerable Earth-like planets--life forms which may have also evolved a human-like consciousness.

Richard Hoover (2011), for example, has presented evidence of ancient bacterial microfossils resembling cyanobacteria in 3 separate meteorites; the remains of organisms which dwelled on astral parent bodies which may have included moons, comets, and planets older than Earth. Hoover also found the remnants of cyanobacteria mats which can take up to 6 months to form. And they were discovered in a meteor older than Earth. It is Cyanobacteria which helped create the oxygen atmosphere of this planet. Cyanobacteria also secrete calcium when creating their mats, and this calcium made it possible for shells, bones, and the skeletal system to evolve (Joseph 2010).

Cyanobacteria are a hardy species, and can live in extreme environments. Therefore, if Cyanobacteria are deposited on Earth-like planets, it can be assumed they would also biologically engineer these alien worlds, providing them with an oxygen atmosphere and flooding the environment with calcium, thereby making it possible for the evolution of bones and brains and for life to evolve into intelligent species, similar to or completely different from, and possibly more intelligent than woman and man.

Most scientists will agree that life on this planet evolved from single celled microbes. Moreover, there is evidence of biological activity in this planet's oldest rocks dated to over 4.2 billion years ago (Nemchin et al., 2008; O'Neil, et al., 2008) when Earth was bombarded with extraterrestrial debris. Therefore, life was present on this planet from the beginning. As there is absolutely no convincing evidence that life began on Earth via an Earthly-abiogenesis, then it seems reasonable to assume that living creatures fell to Earth encased in stellar debris which pounded the Earth for 700 millions years after the creation (Joseph 2000a, 2009a, 2010). Similar events must have taken place on innumerable planets, thereby allowing life to take root and evolve. And if these planets were Earth-like, then life would have also evolved.

And what if these bacterial "seeds of life" fell upon planets unlike our own? If they could take root and flourish, they might evolve into creatures completely unlike those of Earth. This might account for the truly "alien" microbes discovered by Hoover (2011).

In fact, even if we accept that life on Earth began via an Earthly-abiogenesis, then the same must have taken place on a hundred million planets in this galaxy alone.

The implications are profound. It can be assumed that life is everywhere and has a cosmic ancestry which extends backwards in time, interminably into the long ago, and that intelligent life has evolved on countless Earth-like planets (Joseph 2010). And we can predict that life must have continued to evolve on innumerable worlds which are much older than Earth.

But what forms might they take? Might they be human? Intelligent plants? Insects that ask: "are we alone?"

What might be the nature of consciousness on planets with a chemistry and environment completely unlike our own? If complex life exists on worlds billions of years older than Earth, how and in what way might it have evolved? Consider our own planet. If we don't die out, and if science marches on, what might humans be like a million years from now? What about ten million, a hundred million, or a billion years from now? They might seem as "gods" even if they were still human. There are stars which shine in the darkness of night which were born billions of years before the Earth became a twinkle in "god's eye." What might be the nature of consciousness on these ancient worlds?

3. The Conscious Cosmos?

There are tantalizing clues which suggest life may have been present on Earth, fractionating and synthesizing carbon as early as 3.8 to 4.28 billion years ago (Manning et al. 2006; Mojzsis et al. 1996; Nemchin et al. 2008; O'Neil et al. 2008; Rosing, 1999, Rosing and Frei, 2004). Genetic data based on molecular clocks is consistent with these discoveries. Hedges (2001) reports "we found an early time of divergence (approximately 4 billion years ago, Ga) for archaebacteria and the archaebacterial genes in eukaryotes." Hedges (2009) concludes, "life on Earth arose... 4400 to 4200 million years ago, and achieved a prokaryote level of complexity.

There is considerable evidence that the entire eukaryotic genome underwent duplication at the onset of eukaryotic evolution (Makarova et al., 2005). These genes then continued to undergo repeated episodes of single gene and whole genome duplication such that the eukaryotic genome increased in size (Kellis et al., 2004; Dietrich et al., 2004; Dehal and Boore 2005) thereby triggering the transition and divergence between numerous species, ranging from yeast and fungi (Liti and Louis, 2005) to chordates and non-chordates (Dehal and Boore 2005). This would also mean that whole genome duplications and increases in gene numbers led to archae and bacteria diverging from prokaryotes, which was preceded by prokaryotes diverging from proto-cells.

Joseph and Wickramasinghe (2011) determined that the simplest of free-living cells requires no fewer than 382 genes to survive. Therefore, to acquire a genome of this size, beginning with a single gene, would require at least 9 to 10 duplicative events and based on the most liberal of estimate this would have taken at least 1 billion to 2.5 billion years. However, based on estimates as to when life appeared on Earth, and the divergence time between species, i.e. prokaryote from proto-cell, archae and bacteria from prokaryotes, single celled eukaryotes from archae/bacteria, multi-cellular eukaryotes from single celled eukaryotes, Joseph and Wickramasinghe (2011) determined that the first gene was fashioned 10.4<14.5 billion years ago.

The "Big Bang" is estimated to have taken place approximately 10 to 14 billion years ago (bya), with data from the WMAP satellite (based on detailed analysis of the cosmic microwave background) providing evidence for a date of 13.7 bya. This later date has been accepted by current consensus. This range of birth dates for this universe is remarkably consistent with the data obtained from two of the genetic analyses reported by Joseph and Wickramasinghe (2011) i.e. 10.4<14.5 bya.

To speculate: if there was a Big Bang beginning to this universe, then the data reported by Joseph and Wickramasinghe (2011) could be interpreted to mean that the genetic origins of life can also be traced backwards in time to the "Big Bang". If correct, and if there was a Big Bang, then this creation event not only created this universe, but the conditions and essential elements necessary for life in this universe.

Simply put: it could be argued that following Big Bang nucleosynthesis when the universe had cooled sufficiently to create elemental abundances, that the ensuing nucleogenesis and nucleosynthesis led to the production of elements essential for life including carbon (due to triple collisions of helium-4 nuclei) and then the production of molecules such as purine and pyrimidines and the nucleotides adenine, guanine, cytosine, thymine, and uracil, thereby providing all the essential elements for the construction of DNA and RNA. All this could have taken place within several hundred millions years after the presumed Big Bang creation of this universe.

To speculate further: it could be said that if the Big Bang cosmological model is correct, then the Big Bang not only created the essential elements necessary for life, but a Universe in which that life could dwell--a universe ideally fit for incubating life and for the evolution of consciousness. If the Big Bang model is correct, and the genetic analysis performed by Joseph and Wickramasinghe (2011) is confirmed, the implications are staggering: a living universe which was created for the purpose of becoming conscious.

... and the spiraling universe swirled back,
and coiled 'round on planets' knees and shooting stars to ponder its depths in
the cathedral of human consciousness....
... in the mirror of the sea of human consciousness...
... to peer and reflect upon its soul...
... as mirrored in the rising tides of human consciousness...

4. Life in Other Galaxies

Life is everywhere, throughout our galaxy and the cosmos, and life has evolved on planets much older than our own.

Be it a finite (Big Bang) or infinite universe, life many have achieved life, numerous times, but life need only have been fashioned once to spread throughout the cosmos via dispersal mechanisms of panspermia, e.g., comets, asteroid impact, solar winds, rogue planets. Bacteria, archae, and viruses are the ideal intergalactic messengers and can survive and flourish in most any environment. Therefore, it is highly likely that other planets and moons orbiting other stars in this galaxy are also host to single celled organisms which were generated via abiogenesis, and/or which came to live on these worlds after their ancestors were deposited on these planets encased in planetary debris. And it can be predicted that life on some of these planets has evolved, and that viruses and prokaryotes carrying copies of these highly evolved genes have been dispersed to other planets, solar systems, and galaxies thanks to meteor impact, solar winds, and even colliding galaxies (Joseph 2000, 2009a, 2010; Joseph & Schild, 2010). And once deposited on other planets, these microbes may have exchanged genes with the denizens of those worlds, thus influencing the trajectory of evolution on those planets, or if lifeless but located in a habitable zone these microbial sojourners from the stars may have biologically engineered these worlds thus making it possible for more complex species to evolve.

The number of galaxies in the known, Hubble length universe, is unknowable, though if we were to venture a guess, it might be a trillion sextillion. Life could have been independently generated in every galaxy, via abiogenesis taking place in nebular clouds, comets, or those planets with the right ingredients. A single galaxy, such as Andromeda, may contain over a trillion stars (Mould, et al., 2008), each of which is likely ringed with planets. Thus each galaxy likely contains trillions of planets, at least some of which are crawling with life. However, in an infinite, eternal universe, life had to be generated only once and its progeny could then be transferred throughout the cosmos.

Life may be transferred between galaxies when galaxies collide or begin to orbit one another, as is the case with the Sagittarius Dwarf Elliptical Galaxy and the Canis Major Dwarf Galaxy, which orbit and have exchanged stars with the Milky Way (Chou, et al., 2009; Majewski et al., 2003; Martin et al., 2004). Galaxies are in motion and crash into one another from every conceivable direction. It is believed that Andromeda and the Milky Way galaxies will collide in just a few billion years (Cox and Loeb 2008). During collisions, as galaxies merge or as stars (and their planets) are stripped away and exchanged, it is certainly conceivable that life living deep beneath the surface of host planets, would survive galactic transfer. And once in a new galaxy, they could then spread to other solar systems, thereby transferring genes between galaxies.

Horizontal gene transfer (HGT) is common between viruses and prokaryotes, prokaryotes and eukaryotes, and eukaryotes and viruses. Therefore, it can be predicted that genes are shared when organisms from different solar systems and galaxies come in contact (Joseph 2000a, 2010; Joseph & Schild 2010). Through HGT, genetic evolution on one planet, can effect the evolution of species on other planets and even in other galaxies.

Although we can only speculate about the nature of life on other galaxies, we know that carbon-based life equipped with DNA has certainly taken root in the Milky Way. Therefore, it is certainly possible that different galaxies may host DNA/carbon-based life. It is equally likely they also harbor completely alien and unique, non-DNA, non-carbon based life.

It is possible that planets circling stars within the inner arms of the Milky Way galaxy may harbor life forms considerably different from those in the outer arms, where our own sun and planet are located. Therefore, those planets and stars closest to our own would be more likely to host life which is similar to life on Earth as compared to those much further away. And the same may be true of galaxies, with those at the greatest distance from our own being populated with organisms which may become increasingly different with distance. It is conceivable that various corners of the cosmos may be populated by non-cellular or non-DNA derived entities, some of which may also be highly evolved yet in ways that defy human (DNA-based) comprehension. This may be particularly true of life in the most distant galaxies and those whose chemistry may be radically different from our own and which may harbor organisms with a completely different chemistry.

Life on Earth is probably just a sample of life's possibilities.

5. Life, the Habitable Zone and Genetic Engineering of the Biosphere

The nature of life on other planets, at least in our own Milky Way galaxy, likely includes single celled organisms, similar to, if not identical to archae and bacteria, particularly those referred to as extremophiles. Stars give off light and radiation, and rocky planets contain minerals, metals, and a variety of gases, such as hydrogen or even carbon dioxide belched out by volcanoes. Some of these worlds are completely or partially covered with vast amounts of water. Therefore, we can predict that innumerable planets, particularly those located in habitable zones, are crawling with photosynthetic and chemosynthetic organisms and those which feast on acids and metals. We could also venture that these organisms release gases as waste products, such as methane and oxygen. Simple life forms which can extract energy from these "wastes" would also evolve, as well predators which would feast on these microbes, and scavengers which would consume the dead. These creatures would also exchange genes, thereby contributing to evolutionary innovation.

On Earth, horizontal gene exchange is a common currency of exchange among viruses, prokaryote and eukaryotes. The eukaryotic genome was largely shaped by prokaryote and viral genes which triggered multi-cellularity (Joseph 2010). As detailed in this text, these genetic interactions have played a key role in the evolution and diversification of increasingly complex and intelligent species, including modern humans. Other major factors contributing to biological complexity include the environment, the genetically engineered biosphere, and the location of our planet in the habitable zone.

Evolution of complex forms would not have been possible if this planet had been located too close or far away from the sun. Location in the habitable zone, coupled with the rocky watery nature of Earth, also made it possible for the planet's biosphere to be genetically engineered in preparation for those yet to be born. For example, oxygen and calcium had to be manufactured and released in massive quantities, which enabled oxygen breathing creatures with calcium skeletons to evolve and then emerge from the sea, protected by an ozone layer produced by oxygen interacting with sunlight. Thus prokaryotes not only donated genes to eukaryotes, but altered the environment, which triggered the activation of some of these donated genes, thereby promoting the evolution of species perfectly adapted for a world which had been created for them. On Earth, this took over 4 billion years, and once the critical levels of oxygen, calcium, and other elements, had been reached, there was an explosion of complex life such that every modern phyla emerged within a span of less than 40 million years, beginning 540 million years B.P. The biologically altered environment acted on gene expression, and these genes or their antecedents, can be traced backwards in time to viruses and prokaryotes whose ancestry leads to other, more ancient worlds (Joseph 2000a, 2010).

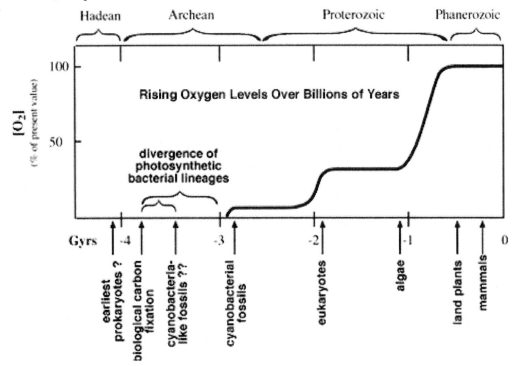

On Earth, the trajectory of evolutionary development continued to enfold over the ensuing 540 million years; a progression which led from boneless fish, to fish, amphibian, reptile, repto-mammals, therapsid, mammal, prosimian primate, monkey, ape, and a dozen or more species of pre-humans. The first species of Homo emerged around 2 million years ago, and via various branching evolutionary trajectories, one branch of Homo evolved into anatomically and neurologically modern humans 30,000 years B.P., whereas others led to evolutionary dead ends, such as Neanderthal.

Given that animal metamorphosis on this planet has been characterized by an obvious step-wise, sometimes leaping linear progression of increasing intelligence, complexity and brain power, it could therefore be predicted that a similar process has occurred on those life bearing worlds similar to our own and which are much older than our own, and those completely alien to our own, such as worlds covered completely with water.

As a progressive increase in intelligence and complexity is characteristic of animal life on this planet, and as up to 90% of the modern human genome is silent and has yet to be expressed (that is, on Earth), it can be predicted that human evolution will continue into the future. Similarly, human-like or animal-like aliens evolving in the older regions of the cosmos would have continued to evolve and undergo metamorphosis, surpassing our own level of development a long time ago. However, it is also likely that these evolutionary-advanced alien life forms, may not be human.

6. Extinction and Evolutionary Possibilities

The path to mammals and humans is just one of genetic evolution's many possibilities. Yet other roads have led to plants, birds, and insects. Although much of the biomass of this planet has not evolved and consists of single celled prokaryotes, the genes donated by prokaryotes (and viruses) to eukaryotes have evolved, and evolution is the hallmark of multi-cellular life on this planet. Life evolves and evolution is a characteristic of life. Thus, there is no reason to conclude that evolution has come to a halt with humans or that evolution does not take place on other worlds.

As extinction and evolution are the nature of life on Earth (Elewa and Joseph 2009; Joseph 2009b), humans may continue to evolve. Or humans may become extinct, and yet other species may take the next evolutionary steps equipped with "human genes" acquired through horizontal gene transfer made possible with the assistance of viruses, archae and bacteria.

Dinosaurs were the dominant species for over 200 million years, but became extinct, paving the way for the ascendency of increasingly intelligent mammals, which led to humans. Every wave of extinction promotes the next stage of evolutionary development, with the progeny of some relatively primitive species becoming much more advanced than those which died out. For example, the egg laying therapsids evolved from repto-mammals and then evolved into and were largely (but not completely) replaced by mammals. Dinosaurs also descended form repto-mammals, and some dinosaurs evolved into birds.

Mammals, insects, and plants also share common ancestors and many genes, including genes contributed by cyanobacteria to the eukaryotic genome nearly 4 billion years ago and probably repeatedly thereafter. Humans, insects, plants, and dinosaurs share common genetic ancestors. If, like the dinosaurs, humans and other mammals also become extinct, evolution would be expected to continue and shared genes might be expressed in plants or insects. Might the Earth of the future be populated with highly evolved intelligent plants with brains, or two-legged insects which gaze into the heavens and wonder: "Are we alone?" Could similar events have taken place on other planets?

7. Genetics and the Nature of Exo-Biological Organisms

It is reasonable to suspect that evolutionary metamorphosis is a principle of life not only on Earth, but on every rocky-watery planet located in the habitable zone of its solar system. Since evolution (and extinction) is characteristic of life on Earth, the same would be true of most every Earth-like planet which harbors microbes capable of biologically modifying the environment.

That does not mean, however, that creatures identical to humans will necessarily "evolve" on every suitable, genetically engineered world. The three domains and five Kingdoms of Life which have taken root on Earth may represent only a partial sample of life and its manifold possibilities as it evolves and struggles for existence elsewhere in the cosmos.

On Earth, the evolution of multi-cellular complexity was made possible via horizontal gene exchange from prokaryotes and viruses, the development of symbiotic relationships where genetically stripped down prokaryotes took up residence within eukaryotes, the insertion of viral genes into the eukaryotic genome, and then complex genetic interactions which altered and repeatedly increased the size of the gene pool and which acted on the environment which acted on gene expression (Joseph 2000a). As is apparent from an examination of the genomes of most species, genes may be shuffled, new genes created, different sections of nucleotides may be suppressed or activated, such that innumerable combinations of genetic elements may be expressed giving rise to diverse traits, characteristics, and species. However, different genetic combinations may prevail on other planets and in environments markedly different from our world. Complex, intelligent species may evolve which are completely unlike anything which has evolved on Earth.

As on Earth, it is likely that some extraterrestrials possess the same "universal" genetic code and a similar genetic endowment consisting of genes made up of nucleotides. Because the three domains and five kingdoms of Earthly life all contain DNA and consist of cellular components, we can make certain predictions about the characteristics of at least some extraterrestrial creatures within our own galaxy and on planets in solar systems closest to Earth. Some alien life forms, like Earth based creatures, consist of living cells. These cells would protect their genes within a nucleus, probably require water, and might have acquired electrical-chemical, generative powers for active transport of food, waste, and the transmission and reception of important messages. Some exobiological organisms would have evolved five or more senses and a brain that could process that information.

Provided a variable Earth-like environment that was susceptible to genetic engineering, we could predict that on certain worlds, over time, a somewhat similar step wise sequence of increasing intelligence, complexity and diversity would take place involving numerous extinctions. Plants and insects, and animals reminiscent of reptiles, repto-mammals, therapsids, mammals, primates, and human-like creatures might likely blossom and unfold; which does not mean they would look like their Earthly counterparts.

8. Alien Minds

There are trillions upon trillions of ancient galaxies consisting of a trillion trillion trillion trillion aged solar systems that are likely ringed with planets -many probably quite like our own. And, just as Life has "evolved" on this world, it could be predicted that Life has emerged on at least a few of these planetary archipelagoes and this would include creatures who long ago "evolved" in a fashion similar to woman and man.

We can predict that those aliens who are genetically related to the Animal Kingdom of Life would be intelligent, and have brains comprised of nerve cells and DNA. Almost all members of the Animal Kingdom, be they vertebrate or invertebrate, have specialized nerve cells and neurons that possess greatly enhanced information processing and intellectual capabilities, as compared to other cells.

Earth based vertebrate and invertebrate brain organization is basically identical at the neuronal level, e.g. possessing cell bodies, dendrites, axons, chemical neurotransmitters and peptides. Further, the genes coding for brain tissues are similar across species and were inherited from common ancestors which were brainless. These genes, or their precursors, did not randomly evolve. Many can be traced backwards in time to the first eukaryotes and prokaryotes whose ancestors hailed from other planets. As prokaryotes and viruses are the ideal intergalactic genetic messengers and can survive in most any environment, then some of this same genetic luggage was likely delivered not just to new Earth, but innumerable other planets.

Hence, those aliens related to the Animal Kingdom of Life and possessing similar genes, would also have evolved brains. Of course, these alien brains may differ from their counterparts on Earth depending on the age, size, climate, and physical-chemical nature of the planet in question.

Animal life on some planets would likely undergo a similar evolutionary sequence in cerebral structural and hierarchical organization and expansion, as is apparent in the progression from fish to primate brain; a process referred to as encephalization. The brain grows larger and more complex. This would include alien brains organized in a fish-like, reptilian, mammalian/primate/human-like fashion. If they have evolved in a manner similar to humans, they may have human-like praxic, creative, technological and linguistic capabilities. If they have evolved beyond humans, then they may have evolved intellectual capabilities which dwarf our own.

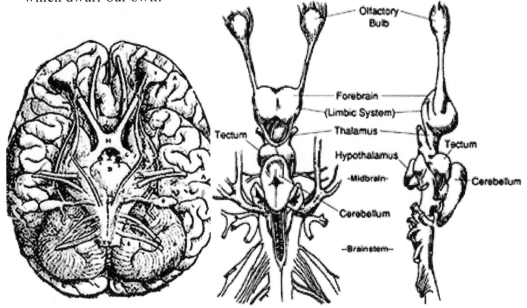

Figure: (Right) Ventral View Human Brain. (Left) Brain 500 mya.

Hence, aliens genetically related to the Animal Kingdom of Life, might posses a midbrain, and brainstem, which would control attention, movement, and sound and visual perception. Their brain would likely include an olfactory forebrain and limbic system (amygdala, hippocampus, hypothalamus, septal nuclei, cingulate) which would mediate all aspects of sex, socialization, love, aggression, memory and emotion, including hunger and thirst. And among the advanced species, they would have evolved a six or seven layered neocortex that would overlay and shroud the older portions of the brain and subserve language and thought and thus the rational mind.

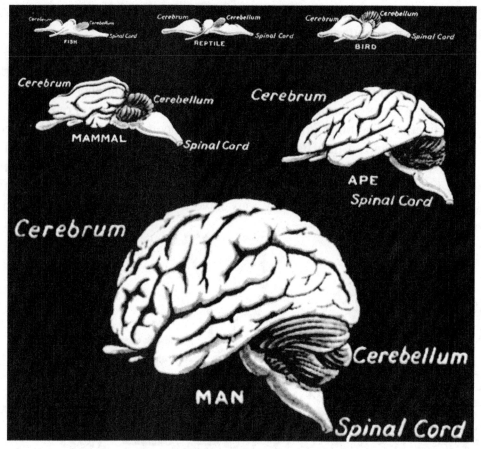

As humans have a propensity for exploration, curiosity and the seeking of knowledge, it could be assumed that those aliens who have reached a similar or advanced level of neurological development would, like Earthlings, share similar social, emotional, and intellectual inclinations, including curiosity, jealously, and pride. Regardless of differences in physical appearance, some advanced alien species -at least those who have evolved from mammal-like creatures, would likely behave, feel, and think in a manner grossly similar to human Earthlings.

For example, like humans, it might be expected that some aliens would thirst for knowledge, or revel in games, sports, competition, conquest, exploitative acquisition, and the glory of rape and slaughter of war. Like modern humans, some aliens would be conquering, enslaving, predatory killers, with little or no compassion for the vanquished. Some aliens might be expected to behave in an irrational and destructive manner, whereas yet others would be seekers of knowledge, wisdom, and spiritual growth.

However, humans or other animals which evolved on planets billions of years older than Earth, may have long ago acquired new brain tissues and new layers of neocortex, and might be capable of processing sensory information which human-Earthling brains may fail to perceive much less comprehend. The humans of Earth have 6 layers of neocortex (which houses the rational, intellectual, logical, analytical regions of the mind). Those who live in the older regions of the cosmos may have evolved 8 or 10, or 12 layers which may be even thicker and contain more neurons than their earthly counterparts.

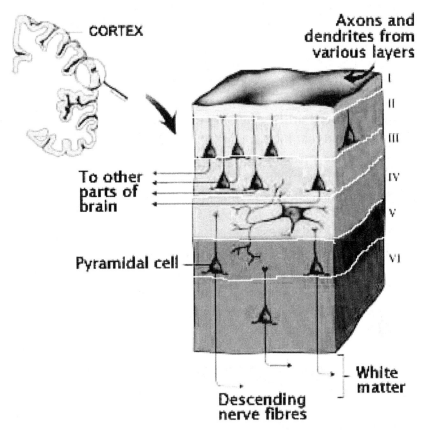

And these evolutionary-advanced extraterrestrials may have experienced expansions in those areas of the brain, such as the frontal lobe which are associated with foresight, judgement, planning skills, and the ability to form long term goals and anticipate long term consequence. These aliens would be correspondingly much more intelligent and creative than any Earthling. If that were the case, they may be able to perceive, deduce, discover, analyze, and assimilate juxtapositions and interrelationships that would be completely beyond human conception.

Alien species need not have evolved on older worlds to have evolved sensory capabilities which dwarf those of human Earthlings. Consider a planet where highly intelligent species have evolved the ability to see via the polarization of light as is the case with insects, and who use echolocation (biosonar), similar to that employed by shrews, bats, and cetaceans. These aliens would be able see objects and animals whose location and shape are defined by sound and would be sensitive to spectrums of light invisible to human eyes. And what if they also evolved the ability to navigate in space by sensing their planet's magnetic fields, as is the case with birds. Coupled that with increased auditory, tactual and olfactory chemosensory perception in a wide range of species. Although the humans of Earth have five senses, there is every reason to suspect that given countless worlds, that some life forms may have developed 10 or more highly developed senses, and a highly developed brain capable of processing this information. There are over a trillion sextillion galaxies in the observable regions of this universe. Each galaxy is likely ringed with a 500 billion to a trillion stars, many with planets like our own. A sextillion sextillion planets in the tiny fragment of the cosmos so far observed. What is the likelihood that only this tiny spec of rock, air and water, our Earth, is the only planet that has evolved intelligent life? What is the likelihood that human intelligence is the crown of creation, and that intellectual evolution stops with us?

9. Extra-Terrestrials: Where Are They?

In 1972 and 1973, NASA attached a pair of gold anodized aluminum plaques to the Pioneer 10 and Pioneer 11 spacecraft, in the hope they would be intercepted by aliens. The plaques featured a nude male and female, along with several symbols designed to tell extraterrestrials about humans and the location of the Earth.

So far, Aliens have not responded.

Where are they? Why haven't they announced their presence?

From the perspective of a hypothetical ET, maybe what has taken place on this planet is so common and mundane, there is no motivation to communicate with a world that is just like a trillion other planets swarming with human-like creatures. Or maybe the humans of Earth are so violent we have been cut off from the rest of the cosmos, isolated, like a disease.

The nature of life, is death, over 99% of all species have become extinct. If the history of the Earth were a 24-hour clock, fully modern humans have been on this planet for only 15 seconds. Unlike other animals, humans flirt with species-wide mass suicide and mutual self-destruction. The human animal can also become extinct. Maybe this is what the future has in store and this is characteristic of human life on other planets.

If human life evolved on other worlds, perhaps they are no more; self-destructing in poisoned planets or the nuclear fires of their own making. Maybe this is a pattern throughout the cosmos: humans evolve and destroy human-life, bequeathing their planets to insects and microbes: the end.

...And before the universe could take a breath, they were no more.

Given our own curiosity as to life elsewhere in the cosmos, it is certainly conceivable that some of those who have evolved in the more ancient corners of the cosmos might be interested in and quite capable of not just observing, but visiting the Earth. If that is the case, it might be asked, then why don't they announce themselves and open up lines of communication?

Perhaps, like good anthropologists they merely observe and gather information, and for the most part leave Earthlings to behave largely unmolested. Speculating wildly, maybe making contact is prohibited and is even against some type of Cosmic Law. Or perhaps there is little or no motivation to visit a world that is just like a trillion other planets swarming with violent, sex-obsessed, power-hungry human-like creatures.

The Search for E.T. relies almost exclusively on the analysis of radio broadcasts. These radio signals are analyzed, for example, by the Megachannel Extra-Terrestrial Assay (META) program. In 2002, Backus and Tarter of SETI believed they'd picked up 11 radio signals, out of 60 trillion, which could not be explained by non-living sources or from Earth. Basing their search and analysis on what they know of human behavior and human technology, SETI decided these 11 signals were probably alien broadcasts. They had all the signatures scientists expect of a human radio transmitter, a frequency that could be transmitted across interstellar space and a very small bandwidth. SETI had discovered humans on another planet? NOPE. Not unless the aliens were playing hide the sausage, because one day it was there, and the next, it was gone. According to the late Carl Sagan these radio signals should be continuous.

All these assumptions are based on a human-centric POV. Sagan, NASA, SETI, all assume aliens behave like humans, think like humans, act like humans, communicate the say way. Problem is, they complain, the universe is getting in the way of finding our brothers and sisters on distant worlds. Interstellar gas clouds and atmospheric turbulence, and a host of other difficulties are believed to be the culprit, when in fact, the problem appears to be with the assumptions of the searchers. Aliens may not be anything like the humans of Earth.

As is apparent from an examination of the genomes of most species, the vast majority of their DNA is silent, the information they contain, unknown. This silent DNA may well have been expressed on other planets giving rise to innumerable species that might never be able to survive on this planet.

Cosmology of Consciousness

It is conceivable that various corners of the cosmos may be populated by non-cellular or non-DNA or non-carbon derived entities, some of which may also be highly evolved yet in ways that defy human (DNA-based) comprehension. This may be particularly true of life in distant galaxies; that is, those whose chemistry is radically different from our own.

10. Do Aliens Listen to the Radio or Watch TV?

There are many who scoff at the idea that life exists anywhere but Earth. For decades the skies have been scoured for extra-terrestrial radio transmissions. The failure to detect them is believed to be evidence that we are truly alone, and that life is present only on Earth which is special and unique.

It is naive, however, to believe that radio or TV transmissions are a yardstick for the measure of intelligent alien life. Highly creative, artistic, talented, and intelligent humans sporting a brain 1/3 larger than modern humans, stalked the Earth over 30,000 years B.P, i.e. the Cro-Magnon peoples (Joseph 2000b). Without benefit of radio, TV, or electricity, Neolithic cultures built Stonehenge, the ancient Egyptians erected the pyramids and in ancient China the "Great Wall." The ancient Babylonians, Persians, Greeks, Romans, Mayas, and peoples of India, and so on, developed science, culture, religion, and created monumental buildings, temples, and empires--all without the benefit of radio, TV or vehicles which could propel them through space.

The very concept of "radio" and wireless communication was not even conceived of until the mid 1880s, and the first successful radio transmissions had to wait until the mid 1890s and only after the creation of the first radio receiver. In the mid 1920s, television was invented. The first space craft to successfully orbit Earth took place in 1957. In the 4.6 billion year history of our planet, these technologies have been around for less than 200 years, and they replaced inferior technologies such as the telegraph and candle lit street lights and homes. Will radio, television, and rocket powered space craft become obsolete in the next 150 years, replaced by some superior technology no one has yet conceived?

There is absolutely no reason to expect that highly intelligent alien life use the same recently invented technology as the humans of Earth; a technology which could become technologically obsolete in another 200 years and which is modeled on and relies upon just two of the human senses, audition and vision. What if alien technology relies upon quantum computers, gravity, magnetic radiation, or sensory capabilities humans have not evolved? The failure to detect electronic transmissions or space craft from other planets does not mean we are "alone."

11. The Evolution of Non-Human Alien Intelligence

On some worlds, highly intelligent species, with an intellect comparable or far superior to humans, may have "evolved" from mammals other than primates, such as wolves and canines, or even insects or plants such that their brains more greatly rely on photo-chemistry or chemo-olfactory-sensations? Dogs and wolves are remarkably like humans, which is why they can live together in harmony. However, there may be "dog stars" where dogs became the dominant life forms. Instead of prosimian primates, it was the ancestors of dogs and wolves which took to the trees. They grew hands instead of claws, climbed back down, stood up, and Dogus Erectus was born! Following a similar sequence as the evolution of humans on Earth, over millions of years Dogus Erectus was replaced by Dogus Neandertailwagus, then Crodogonus, and then, fully modern Sapiens Canis Dogus (the wise dog). There is no reason to assume that dogs, which are so similar to humans, could not have evolved in place of humans, and that on innumerable worlds upright Canis sapiens sapiens stare into the starry night wondering if they are alone.

On some ancient worlds, mammals may never have evolved, such that the metamorphosis of increasingly intelligent life arose from a completely different branch of the forest of life. The direct line leading to hominids may have been killed off, due, perhaps, to that planet's unique environment, climate, atmosphere, gravity, and distance from the sun, and/or because the planet passed through a viral cloud of contagion or was struck by massive debris which chopped off the branch that would lead to primates just before that branch began to bloom.

Consider, on Earth, around 150 mya to 130 mya, flowering plants, flying insects, and transitional species between dinosaur and bird --Archaeopteryx and Anchiornis huxleyi (Elżanowski, 2002; Hou et al., 2009; Yalden, 1984; Xu et al., 2009)-- had begun to coevolve at an accelerated pace, with flying insects and Archaeopteryx/Anchiornis sharing the skies with flying reptiles such as Pterosauria/pterodactyls (Naish & Martill, 2003; Wang et al., 2008), who had taken to wing 220 mya.

Insects, plants and birds, share genes, common ecosystems and a complex interrelationship. Birds and insects feed on flowering/fruiting planets, and both assist in ensuring that pollinating plants breed and continue to reproduce with birds not only spreading pollen but the seeds they eat. Birds feed on insects, and some plants feed on insects, birds, and even small mammals (Barthlott et al., 2007). The physiology of plants and insects show the clearest evidence that both coevolved (Schoonhoven et al., 2006). However, yet another beneficiary of the relationship between insects, plants, and birds, were fruit eating mammals.

Therefore, long before the demise of the dinosaurs, major evolutionary innovations and developments had already taken place, to the mutual advantage of plants, insects, birds, and mammals who began to proliferate. If insects, birds and plants had become extinct with the demise of the dinosaurs, 65 million years ago, mammals would have also likely disappeared from the face of Earth. However, insects, birds and plants would have continued to prevail if mammals along with dinosaurs had been driven to extinction.

The theory of "evolutionary metamorphosis" likens evolution and extinction to the same genetic mechanisms which govern metamorphosis and embryogenesis (Joseph 2000a, 2009b). Just as a "fish-like tadpole" can undergo a complete physical metamorphosis and become a frog, new species may emerge which bear no resemblance to their direct ancestors, except at the level of DNA. For example, some species of dinosaur underwent a complete transformation; they grew feathers and became birds (Hou et al., 2009; Xu et al., 2009). By the same token, if mammals had been driven to extinction, and since birds, insects, and even flowers share many of the same genes as mammals, then in the absence of mammals, what could have evolved in place of mammals could be as different and as unrecognizable as the link between dinosaurs and birds.

Insects and humans also share core sets of genes which can be traced to common ancestors. Therefore, we can predict that if mammals had never evolved or had become extinct, components of the mammalian-human genome would have nevertheless evolved within the genomes of insects who already posses many "human" attributes.

Some insect societies, such as ants, are highly sophisticated, and include divisions of labor, a caste system, altruistic cooperation, complex communication, insect agriculture, the growing of food, weaving, building, the "domestication" of other insects which they corral and milk and eat; and like humans they war against one another and conquer and kill and enslave (Holldobler and Wilson, 1998; Wilson 1974).

Since most of the human and insect genome is "silent" and contains thousands of shared genes which have not yet been expressed, then it is possible that these genes may come to be expressed within the genomes of future species of insects. If humans and other mammals go extinct, social-communal living insects may grow in size, undergo physical transformation and evolutionary quantum leaps, possibly losing their characteristic "insect" physique, and advance to ever heightened levels of communal, cultural, and intellectual achievement. They may only superficially resemble the insects we know today and be as different as birds and dinosaurs, or egg laying mammals vs modern humans.

Likewise, there may be innumerable planets blooming and buzzing with highly evolved intelligent vegetable and insect life; planets where mammals and never evolved or where they became extinct.

Highly intelligent life forms may have evolved on some worlds in place of humans; and like birds they may have evolved a magnetic field sense, and like insects a heightened chemosensory sense, and like dogs a heightened olfactory sense. And they may employ technologies which exploit sensory capabilities that humans lack and would never detect.

On yet other Earth-like worlds, humans may have long ago evolved into something that is no longer human. On planets that are billions of years older than Earth, these alien-humans, like those populated by intelligent plant and insects, may have invented technologies or means of communication completely unlike our own.

What is the likelihood that intelligent plants and insects, or other species of mammal which evolved in place of humans, or evolutionary advanced humans, would be interested in contacting or communicating with the violent, dangerous, waring, raping, mass-murdering, sex-obsessed humans of Earth?

12. Aliens from the Stars

In 1988 story by Ian Watson, "The Flies of Memory" the Earth was invaded by an extraterrestrial race of flies. "We have come to your planet to remember it" they said, and made a request to tour all the Earth's great cities. They were "bloody ugly" creatures who communicated with one another by whistles and chirps. These alien flies had six skinny hairy black legs, the two in the back being the longest which enabled them to walk around on two legs. They were no true flies, not by Earth standards.

They had come from another planet, perhaps another galaxy, where humans had not evolved, or which long ago became extinct. They had come to Earth, to in fact, remember it, and they stared and gazed long and hard at the architecture of our planets greatest structural achievements.

Humans need not evolve on other worlds. Living things completely unlike humans could become the dominant intelligence of billions of planets. Alien life, such as the Flys of Memory, would also process information in ways far superior and completely alien to humans, making them far more superior.

Consider the fly. The flys of Earth have a brain which processes information in a fashion which makes us look quite primitive by comparison. They can process over 100 separate visual impression a second. Humans can muster only a meager 25 discrete images in the same time. The fly's brain enables these creatures to engage in lighting fast maneuverers and make split-second reactions to obstacles, flying around them unerringly, whereas the humans would slam smack dab into whatever was coming his way. For human eyes, anything more than 25 discrete images per second will merge into a continuous movement. A fly can perceive 100 images per second as discrete sense impressions and interpret them quickly enough to steer its movement and precisely determine its position in space.

In the human brain, movements in space produce so-called "optical flux fields" so that objects rush past on the sides, and foreground objects appear to get bigger. Near and distant objects appear to move differently. Fly's perceive movement moment by moment. Therefore, space and time are perceived completely different in the brain of a fly vs a human.

The same can be said when comparing humans to innumerable species. Compare the olfactory sense of dogs vs man. Auditory perception in bats. Creatures which can perceive visual images in ranges of light which are not perceivable to humans.

What of life forms on other planets which have all these superiorities? And why would they be interested in communicate with the humans of Earth, who by comparison might appear to be as primitive as reptiles.

13. THE FUTURE OF EVOLUTION ON EARTH

Yet another possibility speaks to the future of advanced life on Earth which may not include humans. The nature of life, is death, over 99% of all species have become extinct. If the history of the Earth were a clock, fully modern humans have been on this planet for 15 seconds. Species come and go, but not so their genes. Most genes can be traced back to ancestral species which may have served as DNA-hosts, incubating these genes, saving them for transfer and eventual activation. Many of these ancestral genes had not been activated in ancestral hosts. Others underwent only minor modification and were expressed, such as the FOXP2 gene complex which contributes to human speech but exists in the reptile genome. Other strands of DNA underwent single gene or whole genome duplication, freeing them from inhibitory restraint; and once expressed they gave rise to new species.

This raises the possibility that humans may also served as a DNA-host organism, incubating these genes (metaphorically speaking) which may be passed on vertically or horizontally to other species including subsequent species that will evolve in the distant future.

Horizontal gene transfer is common. Humans and numerous species, including insects already share many of the same core genetic machinery.

Thus, when or if humans become extinct, these genes, including those which contribute to "human intelligence" or code for features which are "distinctly human" may become part of the genome of future non-human species which then continue to "evolve."

Moreover, as most of the human genome is silent, coding for functions which are as yet unknown, "silent" genes coding for unknown traits may also be expressed in a future species, much in the same way that FOXP2, which can be found in the fish genome, underwent slight modification and contributed to human speech.

Who or what might be the beneficiaries of the human gene pool if humans cease to exist?

In addition to bacteria, insects are among those most likely to survive if humans destroy themselves in nuclear fires.

Entering the realm of pure speculation, it may be that future species which could be classified as "insects" may continue to evolve and may come to possess not just "human" genes, but some of the characteristics which made humans "distinctly human," such as human intelligence, creativity, and so on, e.g., 6 foot tall "cockroaches" gazing into the heavens a billion years from now, pondering the nature of existence....

14. Genetic Engineering, Designer Babies, and the Eradication of Humans

Humans have acquired the DNA-technology to screen the human fetus for genetic defects, and to chose the sex, eye and hair color of their babies.

If science marches on, and if humans do not self-destruct in a world-wide nuclear war, we can predict that within the next thousand years humans will have acquired and will employ the technology to genetically design babies who are more handsome, beautiful, athletic, and intellectually far superior to their parents.

Naturally, the rich, famous and powerful would be the most likely to afford and the first to employ this genetic-technology. These first generations of genetically altered humans would be members of the privileged class.

It is not unreasonable to assume that these intellectually superior Designer Babies of the future, would develop technologies superior to those of modern humans. Using their greater genetically-enhanced intelligence, these first generation "Designer Babies" may genetically design their own babies who presumably would do the same to their own infants if not to their own bodies and brains.

Therefore, in the next thousand years, a small select group of "humans" may repeatedly genetically engineer their own evolution, and undergo such rapid evolutionary change that they become as different from modern humans as birds are from the dinosaurs. From the perspective of modern humans, the genetically altered humans of the future may become so technologically and intellectually advanced, that they may appear as "gods," whereas modern humans might appear no better than reptiles in comparison.

Human history is replete with mass murder and genocide, with technologically advanced civilizations destroying those which could not compete. Throughout history members of various ethnic groups or "races" have sought to enslave or eradicate other races, which they viewed as inferior "useless eaters."

From the perspective of the first, second, or third generation of designer babies, the bulk of undesigned humanity might appear as inferior, primitive, all consuming competitors for diminishing resources. Therefore, we might predict that the genetically, intellectually enhanced designer babies of the future might decide to completely eradicate and exterminate the last of "modern" humanity, and this is how humans, as we know them, finally become extinct.

As our Milky Way galaxy is over 13 billion years old, and if life in this galaxy began to evolve on Earth-like planets 13 billion years ago, then creatures similar to humans may have also evolved on millions of worlds, in this galaxy alone, billions of years before Earth became a twinkle in "god's" eye. On worlds more ancient than our own, these human-aliens could have also genetically engineered their own evolution, and evolved beyond their counterparts of Earth, long before this planet even formed.

15. The Evolution of the Gods

Can a lizard comprehend a man?
Can a man comprehend a god?
Who dares speak for god?
Perhaps...
Even the Gods have Gods who have Gods...

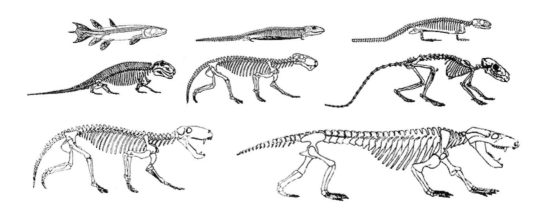

Although interesting theories abound which are supported by experimental and other data, it is unknown if the universe is infinite and eternal and has no beginning and no end, or if it all began with a Big Bang. However, the consensus of scientific opinion is the universe is 13.7 billion years in age. Even if there was a Big Bang there is considerable evidence indicating the universe is much older than 14 billion years. For example, some galactic clusters and walls of galaxies have taken over 100 billion years to form (Lal 2010; Lerner 1990; Mitchell 1999; Van Flandern 2002). It certainly seems reasonable to assume, if there was a big bang, it took place at least 100 billion years ago. Nevertheless, if we were to accept a 13.7 billion year birth date, then coupled with the fact there are galaxies, including the Milky Way, which are over 13 billion years in age (Pace and Pasquini 2004; Pasquini et al., 2005), then it must be recognized that life on planets within this galaxy have also had up to 13 billion years to evolve and to create technologically advanced civilizations. Those who dwell in the more ancient corners of this galaxy may have evolved beyond modern humans 8 billions years before the Earth was formed.

On Earth, the step-wise, branching, quantum-leaping progression which has led from multi-cellular creature to cartilaginous fish, to bony fish... amphibian, reptile, repto-mammal, therapsids, mammal, primate and woman and man, has taken place over the last 540 million years. The species Homo began to evolve 2.5 million years ago. Anatomically modern humans appeared around 35,000 years ago. There is no reason to expect that evolution should stop with modern humans.

Less than 10% of our DNA is necessary for creating a modern day human. Given that over 90% of human DNA is dormant and silent and as tens of thousands of silent genes have yet to be expressed, the likelihood is that human evolutionary metamorphosis on this planet will continue well into the future. Conceivably, these genes will be expressed in the eons to come and give rise to increasingly complex and intelligent human beings. The same could be said of evolution on other planets.

Consider, of the 3-5% of human DNA which is required to build a human, 20% of that is used to create a brain, and another 30% of that DNA is expressed in the process of running and maintaining the brain. That is, 50% of the 3-5% of DNA which is coded, serves the human brain. Since over 90% of the 30,000 or so genes that make up the human genome have not been expressed, this means that thousands of suppressed genes may be available for future cerebral metamorphosis and expansion of the brain.

The brain has increased in size and complexity during the course of animals evolution beginning with a simple forebrain and hindbrain which processed sensory and motivationally significant information and governed movement. These brain areas expanded in size and complexity when animals left the ocean and amphibians then reptiles evolved. However, it may not have been until around 100 million years ago that new cortex (neocortex) began to enshroud the older, more ancient regions of the brain, such that, with the evolution of mammals, six layers of neocortex increasingly covered the older cortex which consisted of 1-3 layers.

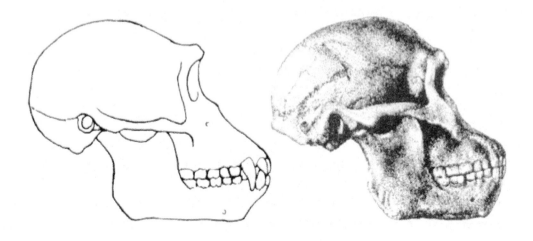

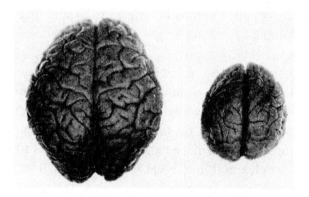

Figure: (Above) Chimpanzee vs Australopithecus Skull. (Right) Human vs Ape Brain

Five million years ago, chimpanzee-like species began to proliferate, including Australopithecus, which had a brain similar in size to that of modern chimps. The brain then tripled in size in the transition from australopithecus, to H. habilis to H. erectus to modern Homo sapiens. However, 30,000 years B.P., the 6ft tall Cro-Magnon people of the Upper Paleolithic had a brain that was 30% larger than modern humans (Joseph, 2000b). And they sported an enlarged frontal lobe which is responsible for insight, foresight, judgment, rational thought, and the ability to set goals, multi-tasks, and to foresee and anticipate multiple possible outcomes. Thus, not only did the brain expand in size, but the modern human brain could easily increase in size and complexity as an older species of humanity in fact sported a much larger brain. A larger, more complex "human" brain may be capable of processing and perceiving information which "modern" brains are only beginning to sense.

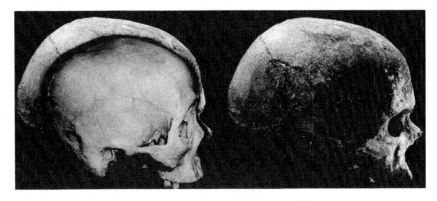

Figure: (Right) Cro-Magnon Skull. (Left) Modern human skull superimposed on Cro-Magnon Skull

Consider, for example, language. Certainly, there is no evidence of reading, writing, or any semblance of grammatical language in human-like species which lived millions of years ago, or as recently as 100,000 years ago (Joseph 2000b). Neanderthals lumbered across this planet for nearly 300,000 years until around 30,000 years ago. Not only was their brain functionally incapable of producing modern human speech, but whatever language they had was likely limited to a few words and emotional sounds denoting a very limited repertoire of concepts and motivational states (Joseph 2000a,b). Language was primitive at best even 100,000 years ago and took several million years to evolve from the sounds produced by H. habilis and australopithecus to the speech of modern humans (Joseph 2000a,b). Thus, whereas the evidence indicates Cro-Magnon people possessed modern language, their contemporaries, the Neanderthals did not. The same evolutionary principles may apply to what is called telepathy and clairvoyance (the so called "sixth sense").

Numerous studies have demonstrated statistically significant evidence of extrasensory perceptual (ESP) ability in some people (Bauer 1984; Beb & Honorton 1994; Radin 2001). Statistically, there is evidence of a significant trend, though very small, indicating that a significant number of people possess very rudimentary psychic abilities. If considered from an evolutionary perspective, ESP in modern humans might be at the level that "modern" language was in ancient humans. The rudimentary language of Australopithecus, H. Habilis, H. Erectus, and Neanderthal consisted of grunts and emotional sounds. This rudimentary language evolved into modern language and the ability to read and write. Likewise, ESP in "modern humans" may evolve into a fully developed 6th, 7th, and 8th sense a million years from now. ESP in modern humans is very primitive and is yet to evolve, just as language was very primitive and rudimentary a million years ago.

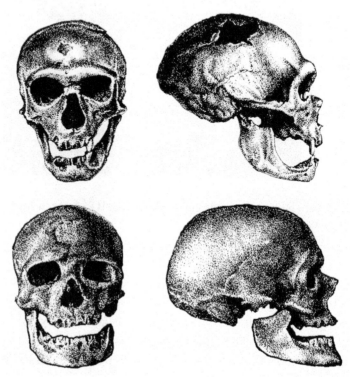

Figure: Neanderthal Skulls. (Bottom) Cro-Magnon Skulls.

However, whereas Neanderthals shared the planet for at least 10,000 years with the more evolutionary advanced Cro-Magnon (who stood 7 inches taller and had a much larger brain), modern humans also likely share the cosmos with evolutionary advanced beings. What might be the perceptual and intellectual abilities of beings which began to evolve before Earth was formed? Might language, the internet, TV, radio, and other primitive means of communication have been replaced by telepathy, clairvoyance, psychic teleportation, and mind reading in those who long evolved in the more ancient corners of the cosmos?

There are trillions upon trillions of ancient galaxies consisting of a trillion trillion trillion trillion aged solar systems that are likely ringed with planets -many probably quite like our own. And, just as Life has "evolved" on this world, it could be predicted that Life has emerged on at least a few of these planetary archipelagoes and this would include creatures who long ago "evolved" in a fashion similar to woman and man. Likewise, since increased complexity and progressive cerebral encephalization is characteristic of the evolution of brain-based life on this planet, the same could be expected on other complex-life bearing worlds. On planets where animals have evolved, it can be predicted that the brain has also increased in size and complexity, giving rise to increasingly intelligent species.

Given what has taken place on this planet over the course of the last 500 million years and the human-genetic potential which is yet to be realized, if similar DNA-based life forms exists elsewhere, then those who dwell in the more ancient corners of the cosmos may have evolved beyond modern humans ten billions of years before the first upright human emerged from the mists of time...and billions of years before the Earth became a twinkle in God's eye....

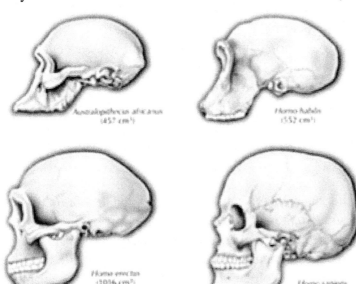

Given evidence of increased brain size in previous species of humanity, coupled with the as yet untapped genetic potential, then it can be predicted that those creatures evolving in the more ancient corners of the cosmos may have evolved additional nuclei and layers of neocortex, exceeding the 6 layered neocortex that is characteristic of mammals and primates--and this may have transpired billions of years before the creation of Earth. In fact, given the unknown antiquity of the universe coupled with tens of billions of years of evolution, the brains of some exobiological organisms may consist of 8, 10, 12 or more layers of neocortex and greatly increased neocortical, perceptual, intellectual, memory, and sensory capacity, thus completely dwarfing the human brain in intellectual, analytical, sensory, perceptual, and cognitive ability. In fact, these ancient ones may have long ago manipulated their own DNA and genetically engineered their own evolution and purposefully developed cerebral tissues that might enable them to analyze, see relationships, or comprehend and manipulate phenomenon that modern humans cannot even perceive much less comprehend. These ancient ones, would be as "gods" even if they were still human.

The Earth is 4.6 billion years young, and much of our science and technology was invented just a few hundred years ago. If science marches on, consider how technologically sophisticated and scientifically advanced humans might be a thousand years from now. What about 10,000 years from now? Or a million years? What about 100 million years from now? Technologically, they would be like gods -even if they were merely human.

And what are the powers of a "god?" What might be the creative and intellectual potency of a being whose brain and civilization is 10 billion years older than our own, and which has likely engineered their own DNA, who continued to evolve and who has 10 or more layers of neocortex --compared to those 6 we call our own? What of a civilization that has had 10 billion years to seek technological perfection?

Consider, the future of human evolution on this planet. If science marches on, and given advances in genetic engineering, the creation of designer babies, designer babies designing their own superior babies, future humans which genetically engineer their own brains and bodies... and imagine the nature of human life 1000 years from now. What of humans a million years from now? Or a hundred million? Or a billion? Then what of those planets which shine in the darkness of night which were formed nearly 10 billions of years before Earth? If life evolved on these ancient worlds, what might they be capable of as compared to the humans of Earth?

It could be expected that these ancient ones might have evolved new brain tissues and additional layers of gray matter and neocortex... gaining intellectual and perceptual capabilities which provide not just a "6th sense" but a 10th sense, completely eclipsing all aspects of human cognition, perception, and intelligence. From the perspective of those who evolved in the ancient regions of the cosmos, the human brain and mind may seem to be just one small step above a frog or a reptile; which, in many ways it is.

Conversely, the human ability to comprehend the intellectual, scientific and technological accomplishments and capabilities of an alien brain organized in this evolutionary advanced fashion, might be analogous to a lizard's ability to comprehend a man. That is, the mental, intellectual, creative, and technological capabilities of creatures who began to evolve ten billion years ago, would lie well beyond human understanding. From the perspective of modern humans, these evolutionary advanced alien humans might seem as gods, even if they were still human.

Archae, bacteria and viruses may serve as intergalactic genetic messengers, influencing the evolution of life on other planets. The ancestors of life on Earth, and their genes, most likely came from older, more ancient worlds. The genetic seeds of life flow throughout the cosmos, and we have likely inherited genes from all manner of life that evolved long before the Earth was formed, including from beings so evolutionarily and technologically advanced, they might seem as gods. As we are the genetic beneficiaries of this cosmic ancestry, this would mean that we are descended from the "gods" and that the humans of the future have the evolutionary and genetic potential to be as "gods" and perhaps evolve beyond good and evil.... perhaps even evolving a cosmic consciousness and *reestablishing* a cosmic unity of singularity which some believe gave birth to this universe in a Big Bang: a consciousness giving birth to itself.

In the unknown antiquity of the cosmos, where life may extend backwards in time interminably into the long ago, even the "gods" may have "gods" who have "gods" who have "gods"....gods which achieve cosmic consciousness... gods who achieve quantum consciousness, gods which become one with the universe... an evolutionary progression which continues eternally, infinitely, and forever into the ever more....

REFERENCES

Barthlott, W., Porembski, S., Seine, R., & Theisen, I. (2007). The Curious World of Carnivorous Plants. Timber Press.

Chou, M-I. et al., (2009). A Two Micron All-Sky Survey View of the Sagittarius Dwarf Galaxy. http://arxiv.org/pdf/

Bauer, E. (1984). Criticism and Controversy in Parapsychology - An Overview.

Cosmology of Consciousness

European Journal of Parapsychology, 5, 141-166.

Bem, D. J., and Honorton, C. (1994). Does Psi Exist? Replicable Evidence for an Anomalous Process of Information Transfer . Psychological Bulletin, 115, 4-18.

Burchella, M. J., Manna, J., Bunch, A. W., Brandob, P. F. B. (2001). Survivability of bacteria in hypervelocity impact, Icarus. 154, 545-547.

Burchell, J. R. Mann, J., Bunch, A. W. (2004). Survival of bacteria and spores under extreme shock pressures, Monthly Notices of the Royal Astronomical Society, 352, 1273-1278.

Chou, M-I. et al., (2009). A Two Micron All-Sky Survey View of the Sagittarius Dwarf Galaxy. http://arxiv.org/pdf/0911.4364.

Claus, G., Nagy, B. (1961) A Microbiological Examination of Some Carbonaceous Chondrites. Nature 192, 594 - 596.

Cox, T. J., and Loeb, A. (2008). The Collision Between The Milky Way And Andromeda, Monthly Notices of the Royal Astronomical Society, 386, 461474.

Elewa, A. M. T. and Joseph, R. (2009). The history, origins, and causes of mass extinctions. Journal of Cosmology, 2009, 2, 201-220.

Elżanowski, A. (2002): Archaeopterygidae (Upper Jurassic of Germany). In: Chiappe, L. M. & Witmer, L. M (eds.), Mesozoic Birds: Above the Heads of Dinosaurs: 129-159. University of California Press, Berkeley.

Flory, D. A., and Simoneit, B. R. (1972). Terrestrial contamination in Apollo lunar samples. Origins of Life and Evolution of Biospheres, 3, 457-468.

Holldobler, B., & Wilson, E. O. (1998) Journey to the Ants: A Story of Scientific Exploration. Belknap Press.

Hoover, R.B., (1998). Meteorites, Microfossils, and Exobiology in Instruments, Methods, and Missions for the Investigation of Extraterrestrial Microorganisms. In Hoover, R. B. Editor, Proceedings of SPIE Vol. 3111, 115-136.

Hoover, R. B., (2006). Comets, carbonaceous meteorites, and the origin of the biosphere. Biogeosciences Discussions, 3, 23–70.

Hoover, R. B., Rozanov, A., (2003). Microfossils, biominerals and chemical biomarkers in Meteorites, in: Instruments Methods and Missions for Astrobiology VI, edited by: Hoover, R. B., Rozanov, A. Yu., and Lipps, J. H., Proc. SPIE 4939, 10–27.

Hoover, R.B. (2011) Fossils of Cyanobacteria in CI1 Carbonaceous Meteorites: Implications to Life on Comets, Europa, and Enceladus. Journal of Cosmology 13.

Hou, L., Zhang, L., & Xu, X (2009) A pre-Archaeopteryx troodontid theropod from China with long feathers on the metatarsus, Nature, 461:, 640-643.
Horneck, G., Becker, H., Reitz, G. (1994). Long-term survival of bacterial spores in space. Advances in Space Research, Volume 14, 41-45.

Horneck, G., Eschweiler, U., Reitz, G., Wehner, J., Willimek, R., Strauch, G (1995). Biological responses to space: results of the experiment Exobiological Unit of ERA on EURECA I. Advances in Space Research 16, 105-118.

Horneck, G., et al., (2001). Bacterial spores survive simulated meteorite impact Icarus 149, 285.

Joseph, R. (2000a). Astrobiology, the Origin of Life, and the Death of Darwinism. University Press, San Jose, California.

Joseph, R. (2000b). The evolution of sex differences in language, sexuality, and visual spatial skills. Archives of Sexual Behavior, 29, 35-66.

Joseph, R. (2009a). Life on Earth came from other planets. Journal of Cosmology, 1, 1-50.

Joseph, R. (2009b). Extinction, metamorphosis, evolutionary apoptosis, and genetically programmed species mass death. Journal of Cosmology, 2, 235-255.

Joseph, R. (2010). Life on Earth, Came From Other Planets. Cosmology Science Publishers, Cambridge.

Joseph R. Schild, R. (2010). Biological Cosmology and the Origins of Life in the Universe. Journal of Cosmology, 5, 1040-1090.

Joseph, R., & Wickramasinghe, N. C., Wainwright, M. (2011). Genetics Indicates an extra-terrestrial origin for life: The first gene. Journal of Cosmology, Volume 16.

Klyce, B. (2000). Microorganisms from the Moon. http://www.panspermia.org/zhmur2.htm.

Lal, A. K. Joseph, R. (2010). Big Bang? A Critical Review. Journal of Cosmology, 2010, 6, 1533-1547.

Levin, G. V. and Straat, P.A. (1976) Viking labeled release biology experiment: Interim results. Science 194, 1322-1329.

Lerner, E.J. (1991), The Big Bang Never Happened, Random House, New York. Majewski, S. R. et al., (2003). A 2MASS All-Sky View of the Sagittarius Dwarf Galaxy: I. Morphology of the Sagittarius Core and Tidal Arms. Astrophys.J. 599 (2003) 1082-1115.

Martin, N. F,., et al., (2004)A dwarf galaxy remnant in Canis Major: the fossil of an in-plane accretion onto the Milky Way. Mon.Not.Roy.Astron.Soc.348:12.

Mastrapaa, R.M.E., Glanzbergb, H ., Headc, J.N., Melosha, H.J, Nicholsonb, W.L. (2001). Survival of bacteria exposed to extreme acceleration: implications for panspermia, Earth and Planetary Science Letters 189, 30 1-8.

Mitchell, F. J., & Ellis, W. L. (1971). Surveyor III: Bacterium isolated from lunar retrieved TV camera. In A.A. Levinson (ed.). Proceedings of the second lunar science conference. MIT press, Cambridge.

Mitchel, W. C., (1997). Big Bang theory under fire. Physics Essays, 10, 342-381.

Mould, J., et al., (2008). A Point-Source Survey of M31 with the Spitzer Space Telescope. ApJ 687 230-241.

Nagy, B., Meinschein, W. G. Hennessy, D, J. (1961). Mass-spectroscopic analysis of the Orgueil meteorite: evidence for biogenic hydrocarbons. Annals of the New York Academy of Sciences 93, 25-35.

Nagy, B., Claus, G., Hennessy, D, J. (1962). Organic Particles embedded in Minerals in the Orgueil and Ivuna Carbonaceous Chondrites. Nature 193, 1129 - 1133.

Nagy, B., Fredriksson, K., Kudynowkski, J., Carlson, L. (1963a), Ultra-violet Spectra of Organized Elements. Nature 200, 565 - 566.

Nagy, B., Fredriksson, K., Urey, C., Claus, G., Anderson, C. A., Percy, J. (1963b).

Electron Probe Microanalysis of Organized Elements in the Orgueil Meteorite, Nature 198, 121 - 125.

Naish, D., & Martill, D.M. (2003). Pterosaurs - a successful invasion of prehistoric skies. Biologist, 50, 213-216.

Nemchin, A. A., Whitehouse, M.J., Menneken, M., Geisler, T., Pidgeon, R.T., Wilde, S. A. (2008). A light carbon reservoir recorded in zircon-hosted diamond from the Jack Hills. Nature 454, 92-95.

Nicholson, W. L., Munakata, N., Horneck, G., Melosh, H. J., Setlow, P. (2000). Resistance of Bacillus Endospores to Extreme Terrestrial and Extraterrestrial Environments, Microbiology and Molecular Biology Reviews 64, 548-572.

O'Neil, J., Carlson, R. W., Francis, E., Stevenson, R. K. (2008). Neodymium-142 Evidence for Hadean Mafic Crust Science 321, 1828 - 1831.

Pace, G., and Pasquini, L. (2004) The age-activity-rotation relationship in solar-type stars A&A 426 3 (2004) 1021-1034.

Pasquini. L., et al., (2005) Early star formation in the Galaxy from beryllium and oxygen abundances Astronomy & Astrophysics 436 3,L57-L60.

Pflug, H.D. (1984). Microvesicles in meteorites, a model of pre-biotic evolution. Journal Naturwissenschaften, 71, 531-533.

Radin, D. (2001). The Conscious Universe: The Scientific Truth of Psychic Phenomena, Harper Edge.

Rode, O.D. et al., (1979). Atlas of Photomicrographs of the Surface Structures of Lunar Regolith Particles, Boston: D. Reidel Publishing Co.

Sharov, A.A. (2009). Exponential Increase of Genetic Complexity Supports Extra-Terrestrial Origin of Life. Journal of Cosmology, 1, 63-65. Schoonhoven, L. M., van Loon, J. J. A., & Dicke, M. (2006). Insect-Plant Biology Oxford University Press. Van Flandern, T. C. (2002). The Top 30 Problems with the Big Bang. Meta Research Bulletin 11, 6-13.

Wang, X., Kellner, A.W., Zhou, Z., Campos, Dde, A. (2008). Discovery of a rare arboreal forest-dwelling flying reptile (Pterosauria, Pterodactyloidea) from China. Proc. Natl. Acad. Sci. U.S.A. 105, 1983-1987.

Wickramasinghe, C. (2011). The Biological Big Bang: Panspermia and the Origins of Life. Cosmology Science Publishers, Cambridge.

Wilson, E. O. (1974) The Insect Societies, Belknap Press.

Xu, X., et al., (2009). A new feathered maniraptoran dinosaur fossil that fills a morphological gap in avian origin. Chinese Science Bulletin, 54, 430-435.

Yalden, D.W. (1984). What size was Archaeopteryx? Zoological Journal of the Linnean Society, 82, 177-188.

Zhmur, S. I., Gerasimenko, L. M. (1999). Biomorphic forms in carbonaceous meteorite Alliende and possible ecological system - producer of organic matter hondrites" in Instruments, Methods and Missions for Astrobiology II, RB. Hoover, Editor, Proceedings of SPIE Vol. 3755 p. 48-58.

Zhmur, S. I., Rozanov, A. Yu., Gorlenko, V. M. (1997). Lithified remnants of microorganisms in carbonaceous chondrites, Geochemistry International, 35, 58–60.

CPSIA information can be obtained
at www.ICGtesting.com
Printed in the USA
BVOW11s1249030117

472452BV00004B/121/P